Mineral Processing in Developing Countries

A DISCUSSION OF ECONOMIC, TECHNICAL AND STRUCTURAL FACTORS

**A United Nations Study
prepared by the Department of Technical Co-operation
for Development
United Nations Secretariat**

A Special Issue of Natural Resources Forum

**Published in co-operation with the United Nations
by Graham & Trotman**

Published in 1984 by
Graham & Trotman Ltd.
Sterling House
66 Wilton Road
London SW1V 1DE

in co-operation with the United Nations

ISBN 978-0-86010-500-8   ISBN 978-94-011-8106-8 (eBook)
DOI 10.1007/978-94-011-8106-8

Acknowledgement
Ivor Herbert, Lorraine Ruffing and Stephen Zorn assisted in the preparation of this study.

The views and opinions expressed in these papers are those of the authors and
do not necessarily reflect those of the United Nations.

Contents

Introduction

This study examines the factors which affect the location of mineral processing in developing countries. These can be divided into two broad categories. The first of these encompasses economic and technical elements affecting the basic viability of a project.[1] These include capital, skilled labour, raw materials, complementary inputs, energy, economies of scale, technological change, growth in demand, proximity to export markets and transport costs. The second consists of structural elements including sources of finance and technology, trade and investment and taxation policies.

The desirability of increasing the extent of processing of natural resources in the developing countries has been emphasized in many international declarations. For example, the United Nations General Assembly, the United Nations Conference on Trade and Development (UNCTAD), the United Nations Industrial Development Organization (UNIDO), the Organization for Economic Co-operation and Development (OECD) and the Commission of the European Economic Communities (EEC) have all supported measures to promote local processing of raw materials in developing countries.[2] A statement on this issue, by the 1974 Latin American Conference on Industrialization, made the following points:

"National industrialization policies should stress the objective of increasing the external autonomy of the developing regions and countries, with special attention to the promotion of exports ... (and) also seek to increase the value added to raw materials being processed and exported.

In the light of the foregoing, it is proposed that national industrialization policies should: (a) promote integrated industrialization based on the potential of each country; (b) stimulate by various means the intensive use of natural resources, both through the promotion of employment policies and through the formulation of schemes which favour the processing of available raw materials."[3]

Among the reasons stated by the developing countries for seeking to increase processing of raw materials are: (a) national strategies for industrialization based on the use of local raw materials; (b) the need to lessen dependence on the industrial countries; (c) the development of skills that can also be used in other sectors of the economy; (d) the ability to enjoy more of the economic rents resulting from raw material pro-

TABLE 1
Mineral Processing Chains

Mineral	1. Mining–Beneficiation	2. Smelting	3. Refining	4. Semi-Fabrication
		Stage		
Bauxite	Bauxite ore-dried bauxite	Alumina[a], Aluminium[a] (billets, ingots)		Rods, tubes, sections
Copper ore	Ore-concentrate	Blister	Refined copper (cathodes, billets, ingots, bars)	Rods, tubes, section
Iron ore	Ore-pellets-sinter	Pig iron	Crude steel (ingots)	Billets, blooms slabs

[a] Bauxite is *refined* to alumina which in turn is *smelted* to aluminium.

1

duction; and (e) the opportunity to obtain investment capital which might not otherwise be available.

The term 'mineral processing' embraces all those activities between the production of ore from a mine and the manufacture of a product for final consumption.[4] However, it is impractical for most developing countries to consider processing of minerals beyond the stage of refined metal ingot or basic ferro-alloy, unless a domestic market of sufficient size exists for economic production of semi-fabricated products.[5] Accordingly, this study concentrates on the production of refined copper, aluminium ingot, crude steel, and selected intermediate products. The basic processing chains involved are shown in Table 1.[6]

It is widely believed that mining and mineral processing methods such as the Bayer process for alumina, the Hall–Heroult process for aluminium, pyrometallurgy and hydrometallurgy of copper, and the use of the blast furnace combined with the basic oxygen furnace (B.O.F.) or direct reduction combined with electric arc furnace for steel are rather well known. However, in all projects process modifications must be made to match the quality of local inputs and requirements.

The issues associated with mineral processing in developing countries have been the subject of a number of other reports.[7] This study pays particular attention to energy requirements, the availability and cost of technology and capital costs. Energy is a growing concern of industrializing nations, since mining and mineral processing are energy-intensive.[8] An important question, therefore, is the impact that energy requirements and availability might have on the location of processing facilities in developing countries.

1 Mineral Processing in Developing Countries

Advantages and Disadvantages

1.1 THE PRESENT SITUATION

The most recent worldwide survey of mineral processing[9] by the United Nations Industrial Development Organization showed that, as of 1977, the developing countries as a whole accounted for an average of some 20–30% of mineral processing capacity (Table 2).

1.2 BENEFITS TO BE GAINED FROM LOCAL PROCESSING

1.2.1 RESOURCE-BASED INDUSTRIALIZATION

Increased local processing of natural resources, and especially of minerals, has become an element in developing countries' proposals for a New International Economic Order.[10] In view of the limited success in the developing countries of import-substitution strategies and of the limited number of countries which have been able to pursue export-oriented strategies based on manufacturing or assembly operations, resource-based industrialization strategies have been given increasing attention by developing countries. Two variations of these strategies have been tried. One assumes that more processing and hence more added value from primary product exports will speed development.[11] The other concentrates on the use of agricultural and natural-resource products not primarily for export, but for domestic consumption. This strategy has been attempted in the People's Republic of China and in the Democratic Republic of Korea. In non-centrally planned countries, the approach has sometimes been advocated, but never wholly put into practice.[12] Some countries have been pursuing both strategies.[13] For example, Chile and Venezuela have undertaken more processing of copper and iron ore, both for export and for use in domestic

industries. Many petroleum-producing countries have adopted industrialization plans based on export-oriented refining and the use of natural gas as a feedstock for petrochemical products.[14]

One of the major arguments for increased processing of natural resources, even where a strategy of production for domestic industry is not immediately feasible, concerns the supposed linkage or 'ripple' effects. These are often said to be of two general kinds:

(a) availability of the product, for local use as well as for export, and the stimulation of linked activities, where local processing of a product makes possible primary production of another product (where, for example, the processing of phosphate rock generates a demand for sulphuric acid);

(b) indirect effects outside the primary-product sector itself in the use of infrastructure, supply of equipment, and fiscal impacts.[15]

Processing projects may not all be accompanied by the desired linkage effects. It may be the case, for example, that a small economy cannot support a manufacturing industry based on local raw materials simply because the market involved is too small.

1.2.2 REDUCING DEPENDENCE

Further processing of natural resources is often seen as a means of reducing a country's dependence on other countries.[16] Several aspects of dependence have been identified:

(a) trade dependence, in which developing countries' ability to import desired consumer and capital goods depends on the export of primary products;

(b) financial dependence, in which the production of raw materials and the construction of infrastructure associated with that exploitation

3

TABLE 2
Developing countries' share in mineral processing capacity
(1977)

	Developing countries' share (per cent)[a]		
	Mine production	Intermediate processing[b]	Metal production[c]
Bauxite–aluminium	62	26	13
Copper	53	39	27
Iron–steel	40	—	9
Lead	34	—	24
Nickel[d]	42	—	27
Tin	88	—	72
Zinc	29	—	15

Source: UNIDO, 1980. Mineral Processing in Developing Countries. United Nations, New York.
[a] Excludes the USSR, Eastern Europe, and the centrally-planned economies of Asia.
[b] Intermediate stages are alumina and blister copper. For other metals, either no distinct intermediate stage exists (lead, tin and zinc), the intermediate stage is almost universally associated with final-stage processing (iron and steel), or the intermediate stage is difficult to define with precision because of the variety of processes used (nickel).
[c] Products are aluminium ingot, refined copper, crude steel, refined lead, refined nickel and ferro-nickel, tin metal, and slab zinc.
[d] Mine production includes some products (e.g. nickel matte from Botswana) which are partially processed but would not be marketable without further processing. Metal production reflects nickel content of final products.

have been financed by capital from external sources;

(c) technological dependence, in which capital goods embodying foreign technologies are imported;

(d) managerial dependence;

(e) market dependence.[17]

The impact of strategies that emphasize processing of raw materials is not clear-cut. For example, additional processing may well improve a country's balance of payments by adding value to exports, but at the same time will tend to increase dependence on export earnings to finance the capital goods and other imports required for processing. Similarly, because efficient resource-processing activities are more often large-scale and capital-intensive, their immediate impact may be to deepen the financial dependence of capital-importing countries. Along the same lines, technological and managerial dependence are likely to be increased where a country must import processing technology, equipment and

management as is generally the case in mineral processing. Forward linkages (i.e. into semi-fabricated products or even to the metal stage) may also risk technological obsolescence, since competing producers may develop more efficient technologies. The development of continuous casting in the copper industry illustrates this type of risk.[18] The same kind of risk exists in the extraction of minerals, but may be less acute because extraction technology is changing less rapidly than processing technology.

The impact of further processing on market dependence varies according to the commodity involved and the extent of processing. In some cases (e.g. production of refined copper or aluminium ingot) processing will widen the marketing options of producer countries, since there are many more metal fabricating companies than there are smelters and refiners. Integration into forward processing can also be the means to promote loyalty among consumers by meeting special needs. However, there are long periods when the sale of refined metals is highly competitive at low prices with surpluses ending up in terminal markets. At the same time, the raw material buyers (smelters and refineries) may be paying high premiums to keep their capital-intensive plants operating at reasonable levels to minimize losses.

1.2.3 DEVELOPMENT OF NATIONAL CAPACITY

Mining and mineral processing produce relatively little direct employment, in relation to the capital deployed.

A UNIDO study estimated the amount of capital investment adjusted to a 1980 basis required to create one job in mineral processing as between US$200 000 to more than $1.5 million (Table 3).

Even if it is wished to increase employment in mineral processing industries, there is little scope for substituting labour for capital. Most technological change in processing has been aimed at increasing the efficiency of raw material use. Given the high share of raw material costs and the relatively low share of labour costs in the value of finished metal products, especially in developing countries where wage rates are lower than in the industrialized nations, there is little incentive to focus on labour-capital substitution. The available evidence suggests that, since most innovations in aluminium and copper processing have significantly increased resource recovery rates as well as an increased capital–labour ratio, there is little scope for plant to operate using older technology in order to generate more employment per unit of output.[19] This conclusion is reinforced by recent increases in energy costs, as older processing technologies are considerably more energy-intensive than new processes.

Relatively high productivity, combined with the low share of wages in total costs, have made mining and mineral processing companies willing, in the recent past, to agree to workers' demands for high wages. The creation of such high-wage enclaves in a small developing economy may lead to rural–urban migration, as workers leave their jobs to seek work in the high-wage sector, even if they also run the risk of a considerable period of unemployment.[20]

1.2.4 TRANSFER-PRICE MANIPULATION

Where both mining and mineral processing operations are under the same corporate ownership, processing facilities in the developing country may reduce the company's opportunities to optimize profits and tax payments through transfer price manipulation. In the bauxite–aluminium industry, for example, the six largest transnational corporations account for 66% of worldwide alumina refining capacity and 54% of aluminium smelting capacity.[21] The price paid to bauxite mines by refineries has often been established by the corporation in order to minimize its worldwide tax liability. Similarly, in the case of copper, a significant amount of trade in unpro-

TABLE 3
Employment in mineral processing

Process	Output per man-year (tons)	Capital cost per job (1980 US dollars)[a]
Alumina refining	800	667 000
Aluminium smelting	90	312 000
Copper smelting-refining	140	450 000
Steelmaking	200	210 000
Lead smelting-refining	225	202 000
Nickel processing (sulphides)	150	1 540 000
Tin smelting	20	205 000
Zinc smelting	200	410 000

Source: UNIDO, 1980, Mineral Processing in Developing Countries, p. 76.
[a] Costs adjusted to a 1980 basis by using Marshall and Swift index of mining and milling costs, as published in Chemical Engineering.

cessed and partially-processed material has been conducted between units of large transnational corporations, although the industry has become markedly less concentrated in the past decade.

1.2.5 ECONOMIC RENTS

A producer of refined metal has a wider range of potential customers than does a producer of un-processed or semi-processed material. Most markets where unprocessed minerals are bought are highly concentrated. Non-ferrous metal smelters and steelworks, as indicated below, have important economies of scale. Often their large size dictates the purchase of supplies from several different mines. The limited number of such processing facilities can reasonably be expected to work to the advantage of the processor as pur-chaser. In contrast, refined metals and minerals are bought by a wide variety of industries.

In the case of copper, there are about 20 independent copper smelters which are prepared to buy concentrate from independent mines. Copper concentrate purchase contracts are usually based on the London Metal Exchange price for refined copper, but payments are arrived at only after complicated deductions for smelting and refining charges, and impurities. In contrast, the producers of refined copper in international trade have a great number of potential customers. Trade in refined metal is normally on the basis of a contract which specifies quantities, chemical specifications, delivery and payment terms, usually on the basis of London or New York Metal Exchange prices. If a supplier of refined copper is unable to find a customer, he can always dispose of his product directly on the exchange, an option not open to producers of concentrate.

The above argument does not necessarily imply that refined metal prices would increase as a result of additional processing facilities in the developing countries, but that such development could in principle obtain for the mineral-producing countries some of the gains otherwise accruing to processors in the consuming countries, though such gains may not be very large.[22] Neither may processing through to final metal stage stabil-ize the export earnings of mineral producers, since the major cause of price instability in minerals markets is variation in demand, in re-sponse to business cycle fluctuations.[23] This varia-tion in demand results from changes in the demand for mineral-containing finished consumer goods and capital equipment, and producers, whether of raw materials, primary metals, or finished goods, must react to these changes in demand either by continuing to produce at full capacity and accepting a decline in prices, by cutting back production, or through strategies containing elements of both. There is no evidence to show that these business-cycle induced fluctua-tions are less severe in the case of primary metals than in the case of ores and concentrates.

1.2.6 ACCESS TO CAPITAL

A final reason for a developing country to pursue 'downstream' processing of minerals may be that foreign capital is easier to obtain for such a project than other kinds of investment funds. Certain industrialized nations which are heavily dependent on mineral imports are prepared to provide public funds for programmes which help to assure long-term supplies. For example, the governments of France, the Federal Republic of Germany and Japan have all subsidized their mining and mineral processing companies' foreign investment ventures so long as such projects included a mechanism for export back to the capital-supplying country.[24] Foreign investors in such ventures may be willing, as a condition for obtaining access to the raw materials, to accept host government insistence on the establishment of local processing facilities, especially where there are substantial home-government subsidies and tax concessions involved. However, the bar-gaining advantage which developing countries have as a result of importing nations' desire to assure their own supplies may be considerably less throughout the 1980s than in the mid-1970s at the height of international concern over raw material supplies.[25]

1.3 BARRIERS TO LOCAL PROCESSING OF MINERALS

A good deal of the discussion of mineral process-ing in developing countries has focused on 'barriers' or obstacles to the development of processing facilities.[26] These barriers can be seen as policies which have resulted in a lesser degree of processing in Third World countries than would have resulted from the operation of 'free' market forces. In the literature on this subject, barriers to processing are often classified as either artificial or natural. The artificial barriers are said to include trade-distorting policies introduced by the industrialized countries, restrictive business practices of corporations which have monopoly or monopsony power, and production-distorting policies of the developing countries. The natural barriers to processing, on the other hand, are the underlying economic characteristics of countries which determine whether they have a competitive advantage in particular processing operations.[27] The latter factors are more properly considered in the context of an overall economic analysis of processing prospects and are dealt with in Section 1.4 below. The following paragraphs discuss some of the more commonly cited barriers to processing.

1.3.1 TARIFFS AND OTHER TRADE LIMITATIONS

The tariff structures of the industrialized countries frequently impose higher rates of duty on the import of processed materials than on unprocessed raw materials. In practice, however, it is not clear that many developing countries are significantly affected by this tariff escalation, at least in the initial stages of mineral processing. The combined effect of the tariff-reducing rounds in the General Agreement on Tariffs and Trade (GATT), the Generalized System of Preferences, which most industrial countries apply to Third World imports, and the preferential-access provisions of the Lomé Convention for exports to the European Economic Community have, taken together, largely reduced tariff barriers on imports of processed metal. One recent study of the impact of tariffs on developing country raw materials processing concludes that "the reduction or removal of developed countries' tariffs on processed raw materials originating in developing countries may not, by itself, do much for the level of processing activity in the Third World".[28]

Similarly, while non-tariff barriers to trade, such as quantitative import restrictions imposed by some industrialized countries, can in theory have a deterrent effect on developing-country processing, there seems to be little concrete evidence to show that such non-tariff barriers do have such an effect in practice in the specific case of non-fuel minerals.[29]

1.3.2 NON-ECONOMIC FACTORS

For a number of reasons, transnational corporations in the mineral sector tend to avoid locating processing facilities in developing countries even if economic factors appear to favour such a location. Transnational corporations attempt to reduce risk by diversifying investments and in particular by avoiding a concentration of investment in countries thought to be likely to nationalize industry. Transnational corporations may also be subject to influence from their home governments, whose defence and strategic interests or whose concern with maintaining domestic employment levels may favour expansion or at least maintenance of their home processing capacities. The extent to which these factors limit a producing country's ability to establish processing facilities depends in great measure on the degree of corporate concentration in the particular mineral industry. Company concentration ratios are very high, for example, in aluminium, while the copper industry is considerably less concentrated.[30] Thus, a developing country would stand a better chance of achieving domestic processing arrangements in copper than in aluminium.

One way in which transnational corporations can restrict developing countries' processing potential is through restrictions on exports from developing countries in which the transnational corporations are investors, or in some cases a complete refusal to permit exports.[31] In the case of minerals, however, there is no clear evidence that transnational corporations have used such monopoly power to restrict developing-country exports in the past decade, although worldwide apportionment of markets for metals has been a feature of past cartel arrangements, such as the short-lived market-control arrangements in copper in the late 1880s, 1889–1901 and 1936–39.[32]

A more specific use of monopoly power sometimes cited as a barrier to developing-country mineral processing occurs if shipping line conferences freight rates are high for processed products.[33] Freight rates are normally higher for materials such as copper cathodes or aluminium ingots than for bulk cargoes like copper concentrates, bauxite or alumina, but the difference between per-ton rates at any time appears to reflect the specific conditions in the world market for ships of a specific type (e.g. there may be high demand for medium-sized freighters which can carry the refined metal, while at the same time there may be a worldwide over-supply of bulk carriers, leading to low freight rates for bulk cargoes). There does not appear to be significant evidence showing that the differences in ocean freight rates can be attributed to shipowners' monopoly power.

A final aspect of monopoly power sometimes cited as a barrier to developing-country processing is the use of massive advertising by transnational corporations to create loyalty not based on product quality differences.[34] In the case of minerals, however, advertising is of comparatively little importance, since quality standards for refined metals are normally set by the various national materials testing organizations and by the metals exchanges themselves. Once a producer has had its brand certified as good for delivery on the relevant metal exchange, little further assurance of quality is required. Some customers may prefer deliveries from industrial-country suppliers, either on the ground that transportation and delivery is more reliable or on the basis of a preference for a particular brand for specific end-uses (e.g. the use of the high-silver fire-refined copper for certain electrical applications), but it appears unlikely that mere advertising would significantly affect buyer preference.

1.3.3   MARKETING PROBLEMS

Selling mineral products requires a reasonably extensive marketing organization; this can be supplied by a state enterprise in the mining and processing industry in a developing country itself, perhaps using foreign firms as agents in specific geographical markets (the strategy followed, for example, by Codelco, Chile's state-owned copper company), or by a foreign investor directly. In any event, marketing will involve costs, such as travel, to establish and maintain sales and distribution outlets and agency arrangements, negotiation of shipping insurance and documentation, and after-sales servicing of customers. The level of such costs can be such that, even though a developing country might have a competitive cost advantage in the actual production of minerals (including transport costs), it may not be able to achieve market entry in some or all of its potential

markets because of the dominance of the marketing costs. The marketing problem is likely to be more severe in the case of metals like aluminium, where markets are highly concentrated, and where the market of last resort – the metal exchange – is not a significant feature, but all producers of refined metals usually see a need for some marketing effort. There is, however, a worldwide shortage of experienced marketing personnel. A number of developing countries have dealt with this problem by selling to merchants rather than consumers, with mixed results.

Related to the marketing issue is the question of communication costs as a barrier to developing-country processing. In the production of fabricated and semi-fabricated metal products, customers' specifications and requirements may change, and the costs of meeting those specifications and the costs of delays can be considerable. Processing locations close to the point of consumption can anticipate changes and reduce such delays and hence reduce total processing-related costs, even though such locations may have higher direct production costs than other locations further from the ultimate markets. Where products are standardized, as in the case of metal ingots, such costs are of only minor importance.

1.3.4   EFFECTS OF TECHNOLOGY

The lack of industrial technology is often cited as a barrier to further industrialization in developing countries.[35] In the case of basic mineral smelting and refining, however, the fundamental technology is widely available from a variety of companies in the industrial nations, and no case is known in which a developing country was unable to purchase or license the required technology, provided the country had access to adequate financing. However, lack of management experience, lack of knowledge of industrial operations, and lack of group know-how are likely to be additional constraints on the ability of developing countries to process their raw materials.[36]

The technological changes occurring in some mineral industries can assist developing countries to establish processing operations. Developments (such as the use of direct-reduction electric furnaces for steel-making) permit the construction of plant on a much smaller scale than was previously economical, thus opening the way for processing for domestic markets in developing countries.[37] On the other hand, other new developments, such as the advent of continuous casting in the copper industry, have the effect of making it more difficult for producers located at considerable distances from major markets to compete effectively.[38] In copper, continuous casting of rod results in a much higher quality wire-making material, which commands a premium reflecting the additional operating and capital costs. Because careful quality control is essential, and because transportation may result in damage to cast coils, rod casting facilities are usually located near their market; thus the spread of this particular technology can be expected to pull copper rod casting away from developing countries.

1.3.5 ECONOMIES OF SCALE

There are efficiencies in mineral processing such that the choice of a plant size smaller than the 'optimum' will be likely to lead to higher costs.[39] In the case of developing countries, this requirement for optimum-size plants tends to be reinforced by the requirement for infrastructure development if any processing at all is going to occur, and by the real economies of scale in certain infrastructure developments, such as hydroelectric facilities. The apparent advantages in constructing an optimum-size plant are often, however, not realized in developing countries. Among the specific difficulties which often arise in such projects are the following:[40]

(a) large plants often experience longer construction times, higher costs, and greater difficulties in arranging utilities, ancillary facilities and infrastructure than do small plants;

(b) large plants tend to experience more technical operating problems than small plants, maintenance may be more problematic;

(c) operating rates tend to be lower in large plants than in smaller plants, thus increasing the ratio of average fixed costs.

Unit costs associated with large mineral processing plants in developing countries may be as much as 40% higher than the equivalent costs if the same plant, with the same factor availability (i.e. energy, complementary inputs, labour, etc.) were located in an industrial country.

1.4 ECONOMIC ANALYSIS OF PROCESSING PROJECTS

The 'barriers' to further processing of minerals in developing countries cited above are not so much absolute blocks to processing as they are factors which tend to have a stronger impact on developing-country projects than on those in the industrialized countries. The major factors which determine which mineral processing projects may succeed in any country are: (1) capital costs; (2) transport costs; (3) environmental control and associated costs; (4) energy availability, requirements and costs; (5) availability and cost of complementary inputs; (6) labour costs; (7) infrastructure requirements and costs; and (8) external costs and benefits of processing projects.

1.4.1 CAPITAL COSTS

Capital costs are typically a high proportion of total costs in mineral processing (with the partial exception of aluminium smelting, where energy costs dominate). For most processing industries, capital costs account for 40% or more of total processing costs, and dominate the non-raw-materials share of the cost structure.[41]

It is not entirely clear whether the dominance of capital costs in processing favours or hampers developing countries. On the one hand, capital goods will tend to be cheaper in the industrialized countries, and the poor conditions under which many Third World plants are built tend to inflate capital charges in those locations.[42] On the other hand, the initial cost of buying and preparing sites for new fabrication plants is rising more rapidly in most industrialized countries than in developing countries,[43] and developing countries may also have access to relatively low-cost sources of finance through the World Bank or other public international institutions, or through bilateral assistance arrangements, that have the effect of reducing the impact on a project of high initial capital costs. What is clear, however, is that capital costs are likely to be the most significant single element in determining the feasibility of a proposed processing venture, and that therefore every effort must be made to obtain estimates of costs that are accurate and to assemble a package of financing that minimizes the annualized capital cost to the project.

1.4.2 TRANSPORT COSTS

A reduction in transport costs may favour a location for processing facilities in the country where the mineral is extracted. The transport savings from undertaking at least initial concentration of low-metal content ores near the mine can be decisive, and such concentration is virtually always carried out in the mineral-producing country in the case of copper, nickel and similar ores.

On the other hand, long-distance transport is quite common for materials with slightly higher metal content, such as bauxite (15–25% aluminium content), copper concentrates with 25–30% metal content, or iron ore with 35–65% metal content. The advent of large-tonnage bulk shipping has made transport of these commodities relatively cheap, and hence discouraged further processing in the country where the material is mined if that is some distance from markets.

None the less, there should be some possibility of further transport savings to be achieved through additional processing. For example, in the case of bauxite, if shipping costs are $13 per ton,[44] and processing costs to convert five tons of bauxite to two tons of alumina are $100, then the transport savings from conversion will equal roughly 40% of the total processing costs, a very considerable saving. In the case of copper, if conversion of four tons of concentrate to one ton of blister copper through smelting costs US$440 (i.e. 20 cents per pound), and shipping costs are the same $13 per ton, then the transport cost saving through smelting prior to shipping would theoretically be $39, or 9% of the processing cost. One should note, however, that this latter saving may not be realized in practice, because of a differential in international freight rates which favours bulk materials like bauxite, alumina or copper concentrate, and which imposes higher per-ton charges on materials like metal ingots. There may also be a possibility that shipping services are only available from a limited number of suppliers, expecially in the case of the cargo liner services, for handling refined metal, and that the shippers might enjoy some of the transport savings achieved through processing.[45]

Even if transport cost savings through processing are small in absolute terms, they may provide some competitive advantage, at least as regards shipments to specific markets. For example, in supplying Japanese markets, the Pacific Island countries could combine processing with shorter transport distances to gain some advantage over, for example, African copper suppliers. African suppliers, in turn, might have an advantage in shipping to European markets. This locational advantage would exist even though the overall ratio of transport costs to total production and processing costs is relatively small.[46]

1.4.3 ENVIRONMENTAL COSTS

A factor of fairly recent origin which may work in favour of the establishment of processing facilities in developing countries is that of environmental protection and pollution control. Many mineral-processing activities, such as alumina refining or copper smelting, are potentially highly polluting. In most developed countries, where sensitivity to environmental issues has increased significantly in the past two decades, complex and costly regulations and restrictions have been imposed on these processing operations so as to minimize pollution. In the USA, for example, it has been estimated that pollution control costs in copper smelting may have increased total capital and operating costs by between 30 and 50%.[47]

The situation is potentially different in many developing countries. Only a very few such countries have elaborate environmental legislation. In addition, because developing countries, by definition, do not have as much industry as the developed countries, they have less existing pollution-causing activity to which the effects of mineral processing would be added. Thus, in purely physical terms, there may be a greater pollution absorption capability in developing countries, particularly if the processing facilities can be located in relatively unpopulated areas.

Even assuming, however, that developing countries may wish to utilize this cost advantage, there are certain potential obstacles. First, the industrialized countries, responding to pressure from domestic industries which face increased costs because of pollution control requirements, may impose 'environmental tariffs' on goods from countries where environmental restrictions are not so severe. Such a policy would be consistent with the fairly common 'sweated labour' tariffs imposed by industrial countries on goods from low-wage countries. In the USA, for example, such tariffs have been advocated for some time.

Second, many international financial institutions, which would be involved in virtually all developing country processing projects, may require environmental controls. The World Bank and the regional development banks, for example, have issued a joint statement requiring environmental considerations to be taken into account in any project in which they are involved as lenders.[48] Such policies might lead developing countries to adopt standards that would remove a competitive advantage.

1.4.4 ENERGY COSTS

Where a developing country has access to relatively low-cost energy, that country may have a significant competitive advantage in energy-intensive mineral processing activities. More than half of aluminium smelting costs, and perhaps one-quarter of copper smelting and refining costs, are for energy.[49] Thus, the availability of fixed sources of low-cost energy (hydro-electric potential, small natural gas fields, or geothermal energy) can make metal processing competitive when energy is important in the total cost structure. Table 4 shows the effect of different energy prices on the cost of aluminium ingot.

More detailed information on the role of energy costs in the various processing operations is given in the sections below dealing with specific metals. It can, however, be seen from the figures in Table 4 that the availability of low-cost power would represent a very significant cost advantage and, moreover, that lack of a source of low-cost power, when combined with the other cost disadvantages typically faced by developing countries, might well make it impossible to set up a competitive processing plant.

There is a substantial variation in capital costs for power plants, especially in developing countries. Typical ranges of costs per installed kilowatt of capacity might be, in 1981 dollars, as shown in Table 5.

These costs could increase by up to 40% for plants in particularly unfavourable locations, including many developing-country locations, and by up to $500/kW for coal plants if all available pollution control equipment is installed.

1.4.5 COMPLEMENTARY INPUTS

In addition to energy, a variety of other complementary inputs are usually required in mineral processing. Alumina production, for example, requires caustic soda and lime, while aluminium smelting requires cryolite, aluminium fluoride and calcium fluoride. Copper smelting requires silica, while refining requires sulphuric acid (itself a byproduct of smelting). Depending on the particular process being used, a specific form of energy (e.g. natural gas) may be desirable. In some cases, the location of complementary inputs is the deciding factor in location of processing facilities; in the case of traditional steel-making,

TABLE 4
Energy costs in aluminium smelting (1980)

Power source	Cost per kWh (US cents)	Cost per pound Al (US cents)[a]
Hydroelectric (established– Iceland, Ghana)	0.6–1.4	3.8–8.9
Hydroelectric (new)[b]	0.75–3.0	4.8–19.0
Coal (Australia)	2.1	13.3
Oil (Japan)[c]	4.5	28.6
Oil (new)[c]	6.0–8.0	38.1–50.8

Source: Trade journal reports.
[a] Based on 14 000 kWh per tonne Al.
[b] Based on capital cost of US$500–2000 per kW, 12.5% annual capital charge, 75% availability.
[c] Based on fuel oil at US$35/bbl.

TABLE 5
Capital cost of electricity generating plant

Type of plant	Cost per kW (US$)
Hydro	500–2000
Geothermal	600–1500
Oil/diesel	500–1200
Natural gas	800–1200
Coal	1000–1500
Nuclear	1500–2500

for example, coking coal availability has often been more important than iron ore or energy costs in determining where steelworks would be located.

Most developing countries do not have the various complementary inputs for mineral processing readily available. This means that they will need to import the required materials, and presumably will have to pay a higher price for such imports than the price which facilities in already industrialized countries will have to pay for locally produced materials. Thus, there is an inherent disadvantage faced by processing facilities in small developing countries in respect of complementary input availability.

1.4.6　Labour costs

In view of the relatively low share of labour costs in the total cost structure of mineral processing (see Table 6), it is unlikely that the developing countries in general possess a major competitive advantage as a result of low labour costs. This is particularly true where domestic wage rates, while lower than those in the industrialized countries, are none the less much higher than in such developing-country manufacturing centers as the Republic of Korea, Taiwan or Singapore, and where a significant number of skilled, managerial and technical employees would need to be brought in from other countries, because the requisite skills are not available domestically. In addition, it is well documented that processing facilities in developing countries tend to have higher levels of staffing per ton of capacity than similar facilities in the industrialized nations; this factor alone cancels out much of the potential advantage that would result from low labour costs.

1.4.7　Infrastructure

It is becoming increasingly common for developing countries to insist that the cost of infrastructure required by mining projects be paid for directly by the mining enterprise. This can be accomplished either by having the mining company directly supply the capital for infrastructure development or by the state's constructing the infrastructure facilities, subject to a prior agreement with the mining company under which the latter undertakes to make annual payments sufficient to

TABLE 6
Labour share in processing costs

Material	Approximate share of total cost (%) due to			
	Raw material	Labour	Other[a]	Capital[b]
Aluminium (input in parenthesis)				
Alumina (bauxite)	30	10	12	48
Aluminium ingot				
(alumina)	31	16	21	32
(bauxite)[c]	9	19	25	47
Copper (input in parenthesis)				
Blister (concentrate)	68	6	7	19
Refined				
(blister)	89	3	4	4
(concentrate)[c]	60	8	10	21

Source: Calculated from UNIDO, Mineral Processing in Developing Countries, pp. 124–39.
[a] Includes complementary inputs and energy.
[b] 12.5% annual capital charge (equal to 10.9% internal rate of return over 20 years).
[c] Capital, labour, and other cost shares at previous stage (alumina and blister copper) are included under those headings rather than as part of raw material cost at metal ingot stage.

meet operating costs and to pay off the capital cost, including interest, over an agreed period of time. Naturally, from the point of view of the mining company, such expenses are a deterrent to investment. In the case of mineral processing projects, as opposed to mining, a company will be likely to prefer to establish facilities where major items of infrastructure such as transportation systems, ports, and power supply are already in place, or can be made suitable for the project at minimal cost, as opposed to supplying a full range of infrastructure at a 'greenfields' site.

To the extent that developing-country governments have access to low-cost sources of finance (e.g. through bilateral or multilateral aid arrangements or international agency loans), they may be in a position to provide certain items of infrastructure, while at the same time obtaining facilities which can be used for purposes other than those of the mining and mineral processing project. In general, however, the relative lack of infrastructure in such countries can be expected to act as a disincentive to investment in mineral processing.

1.4.8 EXTERNALITIES

A variety of economic side-effects, or externalities, are important in the analysis of processing projects. For example, one reason why a copper smelter for some developing country mines may be considered uneconomic is the lack of a local market for the by-product sulphuric acid which would be produced by the smelter.[50] Similarly, a number of United States' alumina refineries were built in the state of Louisiana because of the existence of chemical industries which could supply necessary inputs and purchase byproducts of the refining process, even though other factors, such as shipping costs, favoured locating the refineries near the source of bauxite in Jamaica.

In a country with a variety of natural resources and a relatively large domestic market, it may be possible to think of establishing 'territorial production complexes' based on a systems approach, in which a number of facilities are located close together so they can supply each other with necessary inputs.[51] However, the promotion of such massive industrial projects for supplying regional markets, for example, would appear to offer nearly insurmountable problems to most small developing countries.

The argument concerning externalities and economic linkages can, nevertheless, be turned around and used to justify the establishment of processing facilities in mineral-exporting countries in order to stimulate the growth of related industries. The linkage from mining of ore through smelting and refining to fabrication of metal products and finally to capital goods production is one of the basic patterns of successful industrialization. A recent study shows that the basic metals are among the highest-ranking of industrial sectors when measured by the ability of the sector to generate economic linkages and promote

TABLE 7
Capital cost of new alumina refineries and aluminium smelters

	Refineries		Smelters	
	capital cost		capital cost	
	number	($ per ton)	number	($ per ton)
All projects	11	647	28	4547
New plants	6	676	16	4892
Expansions	5	541	12	3152
Developing country				
New projects	4	580	9	6873
Expansions	2	540	2	2750
Industrial country				
New projects	2	959	7	3068
Expansions	3	543	10	3220

Source: Mining Investment 1981. *Engineering and Mining Journal*, January 1981, pp. 59–81.

growth.[52] Once again, however, the linkage argument has somewhat less force in very small economies, where development of a diversified industrialized economy may in any event be impossible simply because of the small size of the economy.

1.4.9   NEW VERSUS EXPANSION PROJECTS

In virtually all cases, expansion of existing processing capacity can be supplied at a lower capital cost than the same amount of new 'greenfields' capacity. Table 7 shows capital costs per annual ton of capacity for alumina refineries and aluminium smelters currently under construction or about to begin construction. As the table indicates, expansion capacity in alumina can be brought onstream at about 20% less than the cost of new capacity, while for smelters, expansion projects have a 35% advantage over new facilities.

Exceptions to this general rule will result if pollution control requirements in the industrial countries are so stringent that they markedly increase expansion costs.

1.4.10   CONCLUDING REMARK

The various factors reviewed in preceding sections interact in such a complex fashion that it is essential to undertake a well thought-out and realistic economic evaluation of any specific processing venture being considered — an evaluation which quantifies the various factors discussed above, which considers realistic technological options, and which is carried out with a clear understanding of the market situation for the specific commodity involved.

2 Bauxite, Alumina and Aluminium

2.1 THE PRESENT SITUATION

2.1.1 BAUXITE RESERVES

The main ore for the production of aluminium metal is bauxite which is refined to produce an oxide, alumina, which in turn is smelted to produce aluminium. Alternative sources of aluminium-bearing materials are being sought, but there is little likelihood of their displacing bauxite to any significant degree in the near future.

Although bauxite is a relatively abundant material, high-grade deposits are concentrated mostly in developing countries. Guinea, Brazil, Jamaica, India and Cameroon possess over 60% of total world bauxite reserves, and developing countries as a whole account for 74% of reserves. Such concentration has not, however, resulted in any significant market power, because a large part of remaining reserves (about 20% of the world total) is held by one developed market economy, Australia.

2.1.2 BAUXITE PRODUCTION

World bauxite production has grown at an annual rate of 6.2% over the past two decades. While the absolute amount of bauxite mined by developing countries has increased, the rate of growth of such production has only been half that of the developed countries. Consequently, the developing countries' share of world production has decreased from 64% in 1960 to 47% in 1980 (Table 9).

Former major producers, such as Jamaica, Guyana and Suriname, have experienced little growth as compared with the newcomers Brazil, Guinea and Australia. Inadequate port facilities and the efforts of transnational aluminium companies to diversify production in the light of higher taxation imposed by the traditional producing countries have been cited as reasons for this shift.

Six transnational corporations, Alcoa, Kaiser, Alcan, Reynolds, Alusuisse and Pechiney,

TABLE 8
Estimated bauxite reserves (actual weight)

	Millions of tons		Percent of world total	
	a	b	a	b
World	24 600	22 400	100	100
Developing countries	18 180	n.a.	73.9	n.a.
Guinea	8330	5900	33.9	26.3
Brazil	2540	2300	10.3	10.3
Jamaica	1530	2000	6.2	8.9
India	1420	1200	5.8	5.4
Cameroon	1020	n.a.	4.1	n.a.
Guyana	1000	700	4.1	3.1
Venezuela				
Suriname	490	600	2.0	2.7
Sierra Leone	130	n.a.	0.5	n.a.
Developed market economies	5410	n.a.	22.0	n.a.
Australia	4570	4600	18.6	20.5
Centrally-planned economy countries	500	1200	2.0	5.4
Unspecified	510	n.a.	2.1	n.a.

Sources: ᵃ UNCTAD, *The World Market for Bauxite: Characteristics and Trends*, TD/B/IPC/Bauxite/2/Add.1, Geneva, March 1978, p. 52.
ᵇ United States Bureau of Mines, *Mineral Commodity Summaries 1982*, Washington, D.C., January 1982.

TABLE 9
World production of bauxite (aluminium content)
(thousands of tons)

	1960 Volume	(%)	1970 Volume	(%)	1980 Volume	(%)	Annual growth 1960–80 (%)
World	5922.1		12 971.1		19 789.5	100	6.2
Developing countries	3769.1	63.5	7454.2	57.5	9355.3	47.3	4.7
Jamaica	1341.8	22.7	2760.9	21.3	2577.6	13.0	3.3
Guyana	577.2	9.7	1015.4	7.8	652.1	3.3	0.6
Guinea	310.1	5.2	560.3	4.3	2844.0	14.4	11.7
Suriname	691.0	11.7	1204.4	9.3	1047.6	5.3	2.1
Developed market economies	1127.3	19.0	3557.2	27.4	7923.7	40.0	10.1
Australia	15.2	0.3	1990.1	15.3	5806.8	29.3	34.6
Centrally-planned economy countries	1025.6	17.3	1959.6	15.1	2510.5	12.7	4.6

Source: Calculations by the Natural Resources and Energy Division, based on Metallgesellschaft AG, *Metal Statistics 1960–1970 and 1970–1980*, Frankfurt, 1971 and 1981.

control 63% of mine capacity, 66% of alumina refining capacity and 54% of aluminium smelting capacity.[53] Given such dominance, these companies have been able to shift production so as to maximize profits and minimize risk.

Despite the relative stagnation in developing countries' bauxite production, some progress has been made in processing. Between 1960 and 1980, developing country bauxite exports as a share of their domestic bauxite production declined from 85 to 59%, largely as a result of the expansion of refining and smelting capacity in the producing countries.

2.1.3 ALUMINA PRODUCTION

Growth in world alumina production has closely mirrored that for bauxite which was 7% per annum from 1960 to 1980 (Table 10). Developing countries' production increased at twice the rate of developed countries' during the 1960–70 period, doubling their share of world production. That share is currently being eroded, however, by the rapid Australian alumina expansion. In the last ten years, for example, the Caribbean producers have experienced little or no growth in alumina production (Guyana 0.7%, Jamaica 2.9%, Suriname 3.3%). Jamaican output has been re-

TABLE 10
World production of alumina (aluminium content)
(thousands of tons)

	1960 Volume	(%)	1970 Volume	(%)	1980 Volume	(%)	Annual growth 1960–80 (%)
World	4566.1	100	10 598.5	100	17 526.5	100	7.0
Developing countries	496.6	10.9	2186.6	20.6	3638.5	20.8	10.5
Jamaica	338.0		898.7	8.5	1197.5	6.8	6.5
Guyana	0	0	158.5	1.5	148.0	0.8	n.a.
Guinea	85.7	1.9	305.5	2.9	354.0	2.0	7.3
Suriname	0	0	518.0	4.9	720.0	4.1	n.a.
Developed market economies	3142.3	68.8	6543.9	61.8	11 123.0	63.4	6.5
Australia	15.2	0.3	1076.1	10.2	3623.5	20.7	31.5
Centrally-planned economy countries	927.2	20.3	1868.0	17.6	2765.0	15.8	5.6

Source: Calculations by the Natural Resources and Energy Division, based on Metallgesellschaft AG, *Metal Statistics 1960–1970 and 1970–1980*, Frankfurt, 1971 and 1981.

duced in recent years, and no new capacity has been installed, reflecting Jamaica's higher price to United States' markets (see Table 11). Expansions nevertheless, have been announced for Jamaica, as well as for Brazil, India and Australia. Other countries, including Venezuela, Indonesia, Malaysia and Ghana, are studying the possibility of producing alumina. Developing countries' exports as a percentage of their alumina production declined from 87% to 75% during the period, reflecting some expansion of domestic aluminium smelting.

TABLE 11
Average price of United States imports of alumina
(current US $ per ton)

Country	1973[a]	1976[b]	1978[b]
Australia	66	113	139
Guinea	65	n.a.	n.a.
Guyana	65	83	158
Jamaica	70	149	188
Suriname	64	101	163

Sources: [a] United States Bureau of Mines, *Mineral Yearbook 1973*.
[b] United States Department of Commerce, *US General Imports, Schedule a Commodity by Country* FT 135, C.I.F. values (December 1976 and December 1978).

2.1.4 ALUMINIUM PRODUCTION

Worldwide aluminium production grew from 1960 to 1980 at a similar rate (6.5%) to that for bauxite and alumina, but aluminium production in developing countries increased twice as rapidly as in the developed countries though from a low starting point. In 1980, developing countries produced 10.3% of the world's primary aluminium (Table 12). The developed market economies accounted for 69.2% with production being concentrated in the United States (29.0%), Japan (6.8%) and the Federal Republic of Germany (4.5%). These same countries were also the largest consumers (United States, 29.2%; Japan, 10.7%; Federal Republic of Germany, 6.8%) of primary aluminium.

Among the developing countries, India, Brazil, Ghana, Egypt, Suriname, Venezuela and Cameroon produce aluminium. Jamaica, Guinea, Guyana, Indonesia, the Dominican Republic and Malaysia are possible candidates for establishment of smelters. Of these Jamaica, Guinea and Guyana already have alumina refineries which could feed a smelter. In the other bauxite-producing countries, a first step towards greater processing would need to be installation of an alumina plant.

The establishment of further processing can be

TABLE 12
World production of primary aluminium
(thousands of tons)

	1960		1970		1980		Annual growth
	Volume	(%)	Volume	(%)	Volume	(%)	1960–80 (%)
World	4543.2	100	10 257.0	100	16 064.4	100	6.5
Developing countries	113.7	2.5	585.4	5.7	1650.3	10.3	14.3
Brazil	18.2	0.4	56.1	0.5	260.6	1.6	14.2
India	18.2	0.4	161.1	1.6	184.8	1.2	12.3
Ghana	0	0	113.0	1.1	187.7	1.2	n.a.
Egypt	0	0	0	0	120.0	0.7	n.a.
Suriname	0	0	54.9	0.5	54.9	0.3	n.a.
Cameroon	43.9	1.0	52.4	0.5	43.1	0.3	−0.1
Venezuela	0	0.2	22.4	0.3	327.9	2.0	n.a.
Developed market economies	3503.9	77.1	7470.5	72.8	11 128.2	69.2	5.9
Australia	11.8	0.3	205.6	2.0	303.5	1.9	17.6
Centrally-planned economy countries	925.6	20.4	2201.1	21.4	3285.9	20.4	6.5

Source: Calculations by the Natural Resources and Energy Division, based on Metallgesellschaft AG, *Metal Statistics 1960–1970 and 1970–1980*, Frankfurt, 1971 and 1981.

TABLE 13
Level of processing in developing countries

Processing chain	1960	1970	1980
Bauxite production (thousands of tons)	3769.1	7454.2	9355.3
Alumina/bauxite production(%)[a]	13.2	29.3	38.9
Primary aluminium/bauxite production(%)[a]	3.0	7.9	17.6

[a] These figures include primary aluminium production of all developing countries, not merely bauxite producers.

measured by the ratio of bauxite production to alumina and aluminium production. Table 13 sets out such ratios. In 1960, developing countries converted 13% of the bauxite they produced into alumina; this ratio rose to 29% in 1970 and to 39% in 1980. Processing into aluminium was more limited. Developing countries processed 3% of bauxite mined into aluminium in 1960, 7.9% in 1970 and 17.6% in 1980.

TABLE 14
Per capita aluminium consumption

Country	1980 consumption (pounds per capita)
USA	44.1
Federal Republic of Germany	37.5
Australia	33.9
Japan	31.2
Brazil	5.4
Mexico	3.7
Philippines	1.4

Source: Metallgesellschaft AG, Metal Statistics 1970–1980; World Bank, World Development Report 1981.

2.1.5 SEMI-FABRICATION

Information on the production of aluminium and aluminium alloy semi-fabricates and castings is very difficult to obtain for developing countries. According to the World Aluminium Survey 1977,[54] over 40 developing countries possessed some semi-fabricating capacity. Most were producing less capital-intensive aluminium products, such as sheets, circles and extrusions, rather than strip and foil. Total developing country capacity was about 1 million tons, or 9% of world semi-fabricating capacity.

2.1.6 DEMAND FOR ALUMINIUM

Internal demand in developing countries
Developing countries which are considering further processing must evaluate the market for processed products within their own countries. On a global basis, developing country aluminium consumption in 1980 (1 417 000 tons) was slightly less than their production (1 650 000 tons). As per capita income rises, developing countries' consumption levels can be expected to move closer to those of developed countries.

World demand
From 1960 to 1980, world consumption of primary aluminium grew at an annual rate of 7.0%. Principal uses of aluminium are in construction (23%), transportation (23%), packaging (18%), and electrical equipment (11%).[55] Demand growth was particularly vigorous in 1970–73 and 1975–77.

Demand is expected to grow in the future at a slower rate, of perhaps 3.5–5.0% per annum between 1980 and 1990, for a number of reasons. The increase in energy costs may have a paradoxical effect on the demand for aluminium. In the first instance, lower average growth rates in Gross Domestic Product (GDP) caused by higher energy prices, and the general decline in metal consumption relative to GDP in industrial economies, will cause a reduction in the growth of demand for aluminium. Energy conservation, however, should encourage greater use for lighter-weight cars. If aluminium is used in a car in place of steel, for example, there is an immediate weight saving of 118 pounds for each new car. As the aluminium content is increased, fuel savings are achieved.[56] While the transport market may be a major growth area, it is also unreliable, since it is extremely sensitive to recessions.

The very energy-related production costs of aluminium might, however, diminish its competitive edge over other metals such as steel and copper. Energy costs are as much as a third of the total cost of smelting aluminium, compared to less than 10% for other metals. Nevertheless, demand for aluminium is expected to remain strong in at least the container and aerospace industries.

During the 1976–79 period, consumption grew more rapidly than production, so that stocks were depleted and capacity use was increased. The 1975 recession discouraged needed investments, and consequently aluminium capacity was not expected to keep pace with the growth in demand during the 1980s, though demand is still low (1980–82).

The continued growth in demand throughout the 1980s should permit the expansion of smelting capacity in energy-surplus countries. Investors in aluminium production will, of course, show a preference for the location of new plants in those energy-rich countries which also possess the raw materials and which offer favourable terms (especially low power costs) to foreign investors.

The presence of all these requirements has attracted investment to Australia at a rapid rate. Aluminium-making capacity is expected to increase from 325 000 tons to more than one million tons by the mid-1980s though many major projects have been suspended, at least temporarily, recently. Brazil plans to expand its annual capacity from 238 000 to over 800 000 tons. Eventually, other developing countries which have both the raw materials and energy sources can be expected to enjoy an increase in investment in smelting facilities.

2.2 PROCESSING TECHNOLOGY

The principle ore of aluminium is bauxite, a hydrated oxide from which pure alumina (Al_2O_3) is extracted and then electrolyzed to yield the metal. The Bayer process for the extraction of alumina from bauxite was first developed in 1888, and the principles of the Hall–Heroult electrolytic process for recovery of aluminium were discovered in 1886. Despite the passage of almost a century, most aluminium produced today uses the Bayer Hall–Heroult route. So well established are these processes and so extensive are the reserves of bauxite that little change in the process route for producing aluminium is anticipated during the remainder of this century.

As much as 90% of the world's alumina is produced by the Bayer process, using high-grade bauxite as the feed material. The remainder is produced by pyrometallurgical processing of aluminiferous feed material, as for example, in the treatment of nepheline syenite and alunite. In the latter route, outlets for the by-products such as cement and potassium salts are essential, as by-product sales considerably affect the economics of the process.

Aluminium producers, however, have not accepted the conventional routes as necessarily the best and have over the years carried out research into improved and alternative sources and methods. To date, this work has not met with commercial success, but further work is continuing. A review of the alternative processes researched is given below.

2.2.1 ALUMINA PRODUCTION

There are several processes for the production of alumina; among them are the following:

Alkaline processes
In these the aluminium is dissolved as the aluminate. This is the basis of the Bayer process. For low-grade bauxites and other aluminiferous ores, such as nepheline syenites, efforts must be made to avoid the formation of insoluble aluminosilicates. One such method is a process which involves a lime-soda sintering operation.

The Pedersen process
Used for low-grade bauxites, in which the bauxite is smelted with limestone and coke in an electric furnace, the aluminium passing to the slag as an aluminate of lime. The slag is pulverized and treated with a soda solution, so that a solution of sodium aluminate is formed. The solution is then treated with carbon dioxide (flue gases) and the pure aluminium hydrate which is precipitated is calcined and the separated solution re-used to treat additional slag.

Acid leaching
This takes aluminium into solution as Al^{3+} ions. The aluminium salt is crystallized out and the

acids are recycled. Acid leaching is not selective, and iron, in particular, is also leached and contaminates the product. The purification of the crude alumina is therefore necessary, and various process steps have been tried, but thus far without any commercial success. This route is being tested for the treatment of colliery shales and similar aluminiferous materials, but, because of high costs, cannot yet compete with the standard process.

Combined acid and salt leaching processes

Here the aluminium is recovered as a potassium or ammonium compound; the alumina is recovered by calcination. There is little information available on this process, however.

2.2.2 ALUMINIUM PRODUCTION

The Hall–Heroult process is a fused-salt electrolytic process. The principles of the process have changed little in its lifetime, and today's operations differ only in improvements of detail and improved operational efficiency. The major aluminium producers have not readily accepted this technological *status quo*, but have actively researched new processes. These have been mostly based on reduction of the alumina by carbothermal processes but have also included the use of reductants other than carbon. A review of these processes follows.

Carbothermal reduction

When alumina reacts with carbon in an electric furnace, the oxycarbide (Al_4O_4C) is the primary reaction product, and aluminium carbide can then be formed by reaction of the oxycarbide with additional carbon. The carbide can react with alumina to produce aluminium at very high temperatures. The Pechiney company has patented a carbothermal reduction process involving four stages and has had some technical success, but production costs cannot yet compete with those of electrolysis.

Alcoa has investigated a carbothermal process using an electric furnace at a temperature of about 2000°C, and small-scale work has been conducted in Japan on a low-pressure direct-arc reduction of alumina with carbon. However, there is to date no evidence that any carbothermal reduction process is likely to reach commercial operation in the near future.

Use of other reductants

The Toth process is based on the reduction of aluminium chloride by manganese. The process route involves the use of coke, sand, chlorine and manganese. The specific operations include the reduction of manganese oxide, chlorination of clays, purification of the aluminium chloride, and reaction of the aluminium chloride with manganese to form aluminium. The stages involved are complex, and the costs cannot compete with those of the Hall–Heroult route.

To improve the process, Pullman Kellog and Toth Aluminium have announced a two-step approach to combine a number of energy-intensive operations into one process. The work is still at the bench-scale stage; the products will be aluminium-silicon and other alloys.

Purification processes

In the Alcan process, bauxite is first reduced to a liquid ferro-silico-aluminium alloy using carbon in an electric furnace at high temperatures. Then the alloy is reacted with aluminium trichloride at about 1300°C and at atmospheric pressure to produce aluminium monochloride. The volatile monochloride is condensed in a shower of liquid aluminium. Metallic aluminium and the trichloride are formed, the latter being recycled. The high temperature involves substantial energy consumption. Other volatile chlorides can form as impurities and the production of chlorides involves severe corrosion of refractories with resultant high costs.

Other electrolytic processes

Work has been carried out on the electrolysis of aluminium halides and organo-metallic derivatives in non-aqueous solvents. Although energy consumption is low, the basic materials required are very expensive, and both moisture and oxygen have to be excluded.

Electrolytes containing aluminium sulphide have been used in research work. Energy consumption is lower than in the conventional process, but practical problems exist that make the commercial application very difficult. Sulphur evolution will necessitate recovery either for recycling or for sale. Aluminium sulphide in the presence of moisture emits hydrogen sulphide gas which is highly toxic. A closed cell would be required to control this problem, and the construction of such a cell presents practical difficulties.

The electrolysis of baths containing aluminium trichloride ($AlCl_3$) has been investigated by many companies. The latest development in this field and the best known is the Alcoa process, which has reached the semi-commercial stage. There are three process steps involved. First, very pure alumina is reduced. The oxide is then reacted with carbon and chlorine gas to form aluminium chloride. Finally, the aluminium chloride is dissolved in metals of alkali and alkali earth chlorides and the solution is then electrolysed. Liquid aluminium is formed together with chlorine, which is recycled. The process requires careful control for production of water-free $AlCl_3$. The technical aspects of the process require highly skilled supervisors and operators.

This work by Alcoa is of great importance to the aluminium industry, as it may represent a breakthrough in technology. Developments in this technology will be watched with interest mixed with caution. Alcoa, although having built a trichloride pilot plant, will continue to use the Hall–Heroult process in its own expansion.

Despite all the work on alternatives that has been going on over a long period of time, it is evident that the Hall–Heroult process will remain dominant for some time to come. The only competitor on the horizon is the Alcoa process, and that faces great difficulties in obtaining a pure $AlCl_3$ economically. Any new plant to be built in a developing country is likely to be based on proven, well-known and reliable technology, which means for all practical purposes, the Hall–Heroult route.

2.3 THE COST OF PROCESSING

Typical costs in the refining of bauxite to produce alumina and the reduction of alumina to produce aluminium metal are cited below and represent the most recent available estimates for plants in both developed and developing countries. While infrastructure costs (which may be significant in developing-country locations) have not been included, the estimates do incorporate environmental control costs and reflect a variety of possible energy prices.

2.3.1 ALUMINA PRODUCTION

Capital costs
The capital cost of an alumina refining plant is a function of location and a study of recent cost figures indicates that $750 per annual ton capacity (tpy) (1981 prices) is a reasonable estimate for a turnkey plant to produce 500 000 tpy alumina for which the total capital costs would thus be of the order of $375 million. The annual capital charge, based on a 10% interest rate and a repayment period of 15 years, would be of the order of $100 per ton (US¢0.05/lb).

Operating costs
As in the case of capital costs, operating costs, for labour and for consumable process materials, will depend heavily on location. Table 15 indicates representative worldwide costs, in 1981 dollars.

TABLE 15
Operating costs for alumina production
(1981 US dollars per ton alumina)

Input	Price of oil (barrel)[a]		
	$30	*$35*	*$40*
Bauxite: 2.02 tons at $28/ton	56.56	56.56	56.56
Caustic soda: 0.095 tons at $150/ton	14.25	14.25	14.25
Fuel oil: 2.5 barrels	75.00	87.50	100.00
Wages and supervision 2.5 hours at $6/hour	15.00	15.00	15.00
Electricity: 300 kWh at 20 mills/kWh	6.00	6.00	6.00
Miscellaneous	20.00	20.00	20.00
TOTAL	186.81	199.31	211.81

[a] One tonne of fuel oil is approximately equivalent to 6.7 barrels of fuel oil.

Revenue
The selling price of alumina is generally related to the market price for aluminium metal, although each alumina supplier negotiates its own sales arrangements, and there are few published sources of information on alumina prices. As a general rule, an alumina price range of 14–16% of the aluminium price would suggest (given a 1981 United States' producer price for aluminium of US$1675 per ton) a price of the order of $235–270

per ton of alumina. As the cost figures given above indicate, this would appear to make investment in new alumina facilities a marginal proposition at present, although higher aluminium prices could result in a more promising investment outlook. A detailed study of an alumina plant project is always necessary since revenues are very sensitive to variations in input costs.

2.3.2 ALUMINIUM PRODUCTION

Alumina is the feed to the aluminium reduction plant, and is an important element in the cost structure of the aluminium industry. The dominant factor, however, in processing alumina to aluminium is the cost of energy. Energy costs have in the past, and will in the future, dictate the location of aluminium smelters. This is evidenced by the development of aluminium smelting in Canada, Norway, and in the recent years in Australia, Ghana, New Zealand, Venezuela, Brazil and currently in the Middle East, all of which offer low-cost power.

These developments have taken place despite, in some instances, the lack of indigenous supplies of bauxite. Plants have been supplied with imported alumina, for example, in Bahrain, Argentina and the Republic of Korea.

Aluminium is produced from alumina by electrolysis which takes place in a molten bath of natural or synthetic cryolite. The electrolytic reduction cell consists of a carbon-lined box containing a pad of molten aluminium, a carbon anode and the molten cryolite. The anode, which is consumed during the operation is replaced by the Soderberg continuous method or the pre-baked method.[57] The Soderberg technology requires lower investment cost and less highly-skilled workers than the pre-baked system. However, economies of scale, the increased demands of the market and environmental considerations increased the popularity of the pre-baked system. For capacities of 100 000 tpy and over, pre-baked potlines offer the following advantages: (1) lower total investment costs; (2) lower total manpower requirements; (3) lower power consumption; (4) lower carbon and fluoride consumption; (5) easier protection of the environment; and (6) ease of automation and mechanization.[58]

Concerning the environment, it should be mentioned that the potroom atmosphere is worse, and the frequency of lung cancer is increased by a factor of 2.5–3 in Soderberg potrooms, as compared to pre-baked potrooms.[59] It is therefore desirable for new smelters to be of the pre-baked design, and the following cost example is for a plant of that type.

Capital costs
A recently announced aluminium smelter complex is a 100 000 tpy operation to be built in New Zealand. It will be designed so that production can be doubled at a later stage. The capital cost of the first stage of development is estimated to be $4760 per annual ton. This capital cost may be on the high side for a 100 000 tpy plant, as it includes the cost of preparing for further expansion. A capital cost of $4000 per annual ton is more typical, representing a total capital cost of $400 million for a 100 000 tpy plant. A capital charge per ton of aluminium, assuming 15-year amortization and a 10% interest rate, is therfore about $525.

Operating costs
Aluminium production consumes approximately 14 000–16 000 kWh of electricity per ton of product. The lower figure is achieved by most new installations. In Table 16 three different costs of power are used, reflecting different sources of energy: hydropower, coal and oil.

Revenue
Most aluminium is sold on the basis of negotiated contracts; there are, however, reference prices available in the form of North American published producer prices and the London Metal Exchange (LME) settlement price. At the end of 1981, the official North American producer price was approximately $1675 per ton (76 US¢/lb), suggesting that only smelters supplied by low-cost power (2.4¢/kWh or less) would currently be viable.[60]

This situation indicates that smelter development is likely only in those developing countries with hydro-electric resources, unused natural gas, or other sources of cheap and relatively easy to develop power.

2.4 ENERGY USE

The aluminium industry is the most energy-intensive. While mining and concentrating consume barely 2% of the energy used in the industry,

TABLE 16
Costs for aluminium production
(1981 US dollars per ton)

| Input | Power cost (US cents/kWh) | | |
	2.4 (hydro)	5.1 (coal)	8.0 (oil)
1.93 tons alumina at $250/ton	483	483	483
Electricity: 14 000 kWh	336	728	1120
Fuel oil: 0.7 barrels at $35/barrel	25	25	25
Electrolyte: 0.06 tons at $450/ton	27	27	27
Electrodes: 0.6 tons at $150/ton	90	90	90
Labour and supervision: 15 hours at $6/hr.	90	90	90
Miscellaneous	25	25	25
Total operating costs	1076	1412	1804
Capital charge	525	525	525
TOTAL	1601	1937	2329

the production of alumina via the Bayer process consumes 17.5% of the total energy used, and smelting in an electrolytic reduction cell consumes 80.5%.[61]

The data in Table 17 were obtained by Batelle staff from published articles, discussions with producers and private communications. The figure of 244 million British Thermal Units (BTUs) may not represent actual energy use because the figures for electrical energy were all based on a conversion factor of 10 500 BTUs per kWh which is the conversion factor for thermal power plants. However, almost 40% of the electricity consumed by the aluminium industry is generated by hydropower, which has a conversion factor of 3412 BTUs per kWh. Using a weighted average of conversion factors for the different

TABLE 17
Energy use in the production of aluminium

	Millions of BTUs per ton	(%)
Mining	0.29	0.1
Concentrating	4.50	1.8
Alumina	42.60	17.5
Carbon anodes	20.83	8.5
Carbon cathodes	1.21	0.5
Reduction	174.47	71.5
Total	243.90	100.0

Source: Battelle Columbus Laboratories, Energy Use Patterns in Metallurgical and Non-Metallic Mineral Processing, June 1975.

sources of electricity (hydropower, purchased thermal power, self-generated thermal power), a more representative conversion factor would be 8022 BTUs per kWh. This reduces the total energy-use from 244 million to 203 million BTUs per ton of aluminium.

In the years since the completion of the Battelle Institute study, on which Table 17 is based, continuous improvements have been made in the efficiency of potlines. Pechiney, for example, developed 175 000 A pots which cut energy consumption to 13 499 kWh per ton of aluminium, as compared to the 16 000 kWh reported by Battelle.[62] If 14 000 kWh is taken as a representative figure for new facilities, total energy consumption would be reduced from 204 million BTUs to 188 million BTUs per ton of aluminium.

The United States Aluminum Association reports annually on energy consumption within the industry. In 1979 it compared actual consumption with the hypothetical consumption using the best available technology, and its figures are very close to the estimates given in Table 18.

Energy consumption also varies depending on the process route chosen. In the Hall–Heroult process either Soderberg anodes or pre-baked anodes can be used. In both cases the anode raw material is a paste of petroleum and low-ash petroleum coke. The Soderberg technique feeds the paste continuously into the pot, where the heat bakes out the volatiles. The pre-bake system uses natural gas to bake anodes prior to their

TABLE 18

Energy consumption: actual vs. hypothetical, 1979
(millions of BTUs per ton aluminium)

	Actual	Best available technology
Bauxite mining	1.983	1.102
Alumina	38.349	33.060
Smelting	166.842	132.240
Fabrication	33.280	26.448
Total	240.454	192.850

Source: The Aluminum Association, Inc. Energy and the
Aluminum Industry, Washington, D.C., April 1980, p. 20.

use.[63] With power costs ranging from 0.4 to 4
cents per kWh, the total energy cost alone could
vary from $56 to $560 per ton of aluminium.

Soderberg anodes consume approximately
5700 BTU per pound of metal whereas the pre-
baked anodes consume only 1300 BTUs per
pound of metal.[64] It is unlikely that new facilities
will utilize the Soderberg system, not only because
of energy conservation but also because volatiles
can be better controlled in the pre-baked system.

Energy can be saved not only through improvements in the Hall–Heroult process, but also
by increasing the use of scrap. It presently takes
only 13 million BTUs per ton to refine scrap to
molten metal fit for use, a saving of 90% of the
energy costs in smelting. Scrap has grown to 20%
of total aluminium consumption. However, the
existence of scrap is dependent on the production
of primary aluminium, so there is a technical limit
to energy savings via scrap smelting.

Besides fine-tuning the potlines to conserve
energy, there are more radical experiments which
have achieved sizeable reductions in energy con-

sumption. The Alcoa process already described
starts with the Bayer-processed alumina which is
transformed to aluminium chloride and then fed
into an electrolytic cell. Alcoa claims this process
requires 4.5 kWh per pound aluminium as compared to 6.4 kWh under the best available Hall–
Heroult technology. However, this process has
not yet reached commercial viability.

Another consideration is the form in which
energy is used at each stage of processing. Table
19 illustrates the distribution of type of energy
used at the different stages. Electric power costs
can vary significantly, depending on the source of
the power: hydro, coal or oil. An inspection of
recently-published power costs confirms the fact
that among the energy power currently in use by
the industry hydropower is the least expensive
(Table 20).

TABLE 20

Actual electric power costs 1979
(US cents per kWh)

Country	Hydro	Coal	Oil
Canada	0.4	1.1	3.6
Japan[a]	—	—	4.0
USA[b]	0.6	2.5	—
Australia[c]	—	1.6	—

Sources: [a] K. F. Kangas, Energy Outlook for the Western
World Aluminium Industry, Research Securities of
Canada, April 1980, p. 4.
[b] Metal Bulletin, "Aluminium and Energy in the
1980s". October 2, 1980, p. 23.
[c] J. Cook, Australia's Changing Role in the World
Aluminium Industry, Paper delivered at World
Aluminium Congress, Madrid, September 1980.

TABLE 19

Distribution of energy by type (%)

Stage	Electricity	Natural gas	Other fuels	Total
Mining, concentrating	6.0	40.0	54.0	100
Alumina refining	12.0	80.5	7.5	100
Molten metal smelting	85.2	3.2	11.6	100
Hold, cast, melt	7.4	80.0	12.6	100
Fabrication	38.1	50.8	11.1	100
All stages	64.7	24.3	11.0	100

Source: H. W. Lownie, et al, Development and Establishment of Energy Efficiency
Targets for Primary Metal Industry, Battelle Columbus Laboratories, Columbus
1976.

There are several bauxite producers which could increase their export earnings by further processing (see Table 21). Refining bauxite to alumina is less energy intensive than smelting the alumina to aluminium. More important is the fact that the two processes depend on different energy sources: alumina requires a *direct* heating fuel (natural gas or oil) while aluminium uses electricity which can be hydroelectricity. Thus, the ability of a bauxite producer to fill the processing gap may depend not only on the relative quantity of energy but also on the type of energy available.

The energy requirements related to closing the 'processing gap' are summarized in Table 21. The processing gap was estimated in a very general way by ratioing the part of mine production processed within the country, with that exported in crude form. However, production is affected by short-term fluctuations, so that some measure based on capacity is more meaningful. For example, a recent UNIDO study determined the 'processing gap' by calculating the tonnage of processed material that could be produced from full capacity production of a country's mines less any processing capacity already in place.[65] The following table uses this definition of the 'processing gap'. This difference is estimated for each producing country, and compared with minimum processing rates. Those differences smaller than the minimum processing rate have been removed from the list.

Ironically, the largest bauxite owners produce the least energy. Current energy production in Guyana, Suriname, Jamaica, the Dominican Republic, Haiti, Guinea and Sierra Leone is inadequate for increased processing. Guyana, Suriname, Ghana, Guinea, Sierra Leone on the other hand have hydropower potential which

TABLE 21
Energy requirements for bauxite processing

Country	'Alumina gap' (thousands of tons) 1	Energy required (million toe)[b] 2	'Aluminium gap' (thousands of tons) 3	Energy required (thousands of MWh)[b] 4	Primary fossil (million toe) 5	Domestic energy Production nuclear, hydro (million toe)[c] 6	Primary energy fossil fuel (million toe) 7	Reserves hydro potential MW 8
Brazil	1416	0.435	786	11 004	12.281	8.618	5735.850	106 500
Guyana	1825	0.560	1806	15 204				12 000
Suriname	2175	0.668	1675	23 450		0.111		260
Jamaica	4120	1.260	3565	49 910		0.010		
D. Republic	565	0.173	283	3962		0.007		
Guinea	5865	1.800	3285	45 990		0.007		6400
Sierra Leone	305	0.093	150	2100				3000
Ghana	175	0.184			0.020	0.356	1.372	1615
Cameroon						0.109	19.310	22 960
India	315	0.098	200	2800	66.176	3.556	648.530	70 000
Indonesia	650	0.200	325	4550	95.365	0.184	2820.000	30 000
Malaysia	475	0.146	240	3360	9.770	0.087	965.500	1319
Venezuela[a]	1000	0.307	225	3150	124.726	1.067	2133.000	36 000
Philippines[a]			140	1960	0.163	0.436	3.400	7504

Sources: Col. 1, 3: UNIDO, 1978. *Mineral Processing in Developing Countries*, Chapter 3, and other sources.
   Col. 2, 4: Author's estimates.
   Col. 5, 6: UN, 1979. *World Energy Supplies 1973–1978*, NY.
   Col. 7, 8: World Bank, 1980. *Energy in Developing Countries*, Appendices.
[a] The UNIDO study did not estimate a 'gap' for Venezuela and the Philippines since they are not yet producing bauxite. Nevertheless, energy requirements were calculated based on firm plans for expansion.
[b] Energy required to produce one ton alumina = 26.4 million BTU or 0.615 toe.
Brazil 1 416 000 divided by 2 multiplied by 0.615 = 435 999 toe or 0.435 million toe.
Energy required to produce one ton aluminium = 14 000 kWh = 24 MWh
Brazil 786 000 tons × 14 MWh = 11 004 000 MWh.
[c] Primary energy generated from hydroelectric stations is measured under 'ideal' conditions, that is 100% efficiency where 1 kWh equals 3412 BTUs.

might be developed to meet their processing requirements. Cameroon currently smelts aluminium and has large unexploited bauxite deposits (1 billion tons) as well as oil (19 million toe), gas (28 million toe) and hydropower (22 960 MW potential). Thus, the potential to establish a completely integrated complex appears to be good. Ghana, Guinea and Sierra Leone might be able to promote further processing on a regional basis using the hydro potential in Cameroon, Central African Republic and Upper Volta. There would, of course, be both technical and political issues to be resolved.

The Dominican Republic, Jamaica and Suriname are also energy-deficit countries. Their best possibility to increase or to enter processing would seem to be through joint ventures with energy-surplus countries such as Mexico, Venezuela, Guyana and Trinidad. A forerunner of such co-operation is the current Mexican–Jamaican agreement whereby Mexico will sell oil in exchange for Jamaican bauxite. In addition, Mexico will take a 10% share in Jamaica's new alumina plant.

2.5 AVAILABILITY OF ALUMINIUM TECHNOLOGY

In the aluminium industry, the major transnational companies have established themselves as vertically integrated organizations.

The industry outside of the centrally planned economies is dominated by six large corporations. These are: Aluminum Company of America (Alcoa), USA; Pechiney Ugine Kuhlmann (PUK), France; Swiss Aluminium (Alusuisse), Switzerland; Aluminum Company of Canada (Alcan), Canada; Reynolds Metals Company, USA; Kaiser Aluminum and Chemical Corporation, USA.

Technology is available from other sources; an example is the aluminium smelter in Dubai which started production in late 1979 and had an annual capacity of 135 000 tons in 1981. Dubai Aluminium Co. (Dubal) operates a power station and a desalination plant in addition to the smelter. The power station and desalination plant accounted for just over half the total cost. Technology for the smelter and start-up personnel was supplied by National Southwire Aluminum (NSA) of the USA based on that used in their Kentucky smelter. Dubai has recruited experienced personnel from Canada and the United Kingdom as well as from the Bahrain smelter. The decision to build the plant was taken by the Dubai Government in 1975. Gas turbines were chosen for the production of electricity with waste heat from the power station used to produce potable water. The project construction was controlled by British Smelter Constructions who divided the very large project amongst United Kingdom and West German contractors.

Technological expertise can also be found in Germany, South Korea, Japan and Australia. The USSR and Hungary, in particular, have large aluminium industries and assistance from these countries has been provided to some developing countries. Alumina plants have been built in Ireland, Brazil, Turkey as well as in the Caribbean Islands, whilst aluminium plants have been built in Ghana, Venezuela, Suriname, Bahrain and several other developing countries.

3 Copper

3.1 THE PRESENT SITUATION

3.1.1 COPPER RESERVES

Developing countries account for over half of copper reserves in the world (including centrally-planned economies). Among the largest reserve holders are Chile, Peru and Zambia. However, large shares are held by developed countries, namely the United States and Canada (Table 22).

3.1.2 COPPER ORE AND CONCENTRATE PRODUCTION

World production of copper ores and concentrates has been growing at an annual rate of 3.1% during the 1960–80 period. Developing countries mined 43% of the world's copper ores in 1980. Chile, Peru, Philippines, Mexico, Zaire and Zambia are among the oldest producers. They have recently been joined by Indonesia and Papua New Guinea. Zambia and Zaire appear to be losing their relative shares as a result of transport and other difficulties (Table 23).

Exports of ores and concentrates as a percentage of total mine production rose from 9% in 1960 to 22.5% in 1980. This reflects the expansion of mining capacity in Indonesia, the Philippines and Papua New Guinea which was unaccompanied by a parallel increase in processing facilities in those countries.

3.1.3 SMELTING AND REFINING OF COPPER

The operations of the older producers, Chile, Peru, Zaire and Zambia, are largely integrated from mining and concentrating through smelting, refining and some limited semi-fabricating. Smelting near the mine site saves transport costs since 70–75% of the weight of concentrate is

TABLE 22
World copper reserves

	1979[a]		1982	
	(millions of tons)	(%)	(millions of tons)	(%)
World	492.50	100.0	505.0	100.0
Developing countries	264.69	53.7	n.a.	n.a.
Developing countries in North America	29.90	6.1		
Chile	97.00	19.7	97.0	19.2
Peru	31.70	6.4	32.0	6.3
Other developing countries in South America	9.97	2.0	n.a.	n.a.
Zaire	23.08	4.7	30.0	5.9
Zambia	33.55	6.8	34.0	6.7
Other African countries	11.79	2.4	n.a.	n.a.
Asian developing countries	27.20	5.5	n.a.	n.a.
Developed market economies	167.60	34.0	n.a.	n.a.
Europe and Near East	21.70	4.4	n.a.	n.a.
United States of America	91.60	18.6	90.0	17.8
Canada	31.70	6.4	32.0	6.3
Oceania	22.60	4.6	n.a.	n.a.
Centrally-planned economies	59.80	12.1	60.0	11.9

Source: United States Bureau of Mines, *Copper: Mineral Commodity Profiles*, 1979, p. 7 and *Mineral Commodity Summaries 1982*, p. 41, 1982.
[a] The Europe, Near East and Oceania figures contain some reserves from developing countries.

TABLE 23
World production of copper ores and concentrates (Cu content)
(thousands of tons)

	1960		1970		1980		Annual growth 1960–80 (%)
	(Volume)	(%)	(Volume)	(%)	Volume	(%)	
World	4239.2		6383.5		7816.2		3.1
Developing countries	1885.2	44.5	2445.1	38.3	3340.3	42.7	2.9
Chile	532.1	12.6	691.6	10.8	1067.7	13.6	3.6
Indonesia	0		0		56.6	0.7	
Papua New Guinea	0		0		146.8	1.8	
Peru	184.0	4.3	220.2	3.4	365.3	4.6	3.5
Philippines	44.2	1.0	160.3	2.5	304.6	3.8	10.1
Mexico	60.3	1.4	61.0	0.9	175.4	2.2	5.5
Zaire	302.3	7.1	387.1	6.1	459.7	5.8	2.1
Zambia	576.4	13.5	684.1	10.7	595.8	7.6	0.2
Developed market economies	1729.4	40.8	2722.6	42.7	2664.1	34.1	2.2
Centrally-planned countries	624.6	14.7	1215.8	19.0	1811.8	23.2	5.5

Source: Calculations by the UN Natural Resources and Energy Division, based on Metallgesellschaft AG, *Metal Statistics 1960–70 and 1970–80*, Frankfurt, 1971 and 1981.

waste. Consequently, bulk trade in copper concentrates has never been as significant as for iron ore or bauxite.

Production of refined copper has been growing at about the same rate as mine output, 3.2% per year, in the 1960–80 period (Table 24). The share of developing countries in refined production has increased over this time, from about 18% of the world total in 1960 to over 22% in 1980.

One can assess progress in processing by comparing mine, smelter and refinery production over time. Zambia smelts and refines almost everything it produces. Chile, Peru and Zaire could increase somewhat the amount of mine output they smelt or refine. In 1980 Chile smelted 89% of mine output and refined 76%; Peru smelted 89% and refined 63%; Zaire smelted 93% and refined 31%. The Philippines, Indonesia and Papua New Guinea do not, at present, smelt or refine any copper, although a smelter is under construction in the Philippines.

Chile has a long history of copper smelting and

TABLE 24
World production of refined copper (thousands of tons)

	1960		1970		1980		Annual growth 1960–80 (%)
	(Volume)	(%)	(Volume)	(%)	Volume	(%)	
World	4991.1		7582.8		9362.1		3.2
Developing countries	893.3	17.9	1494.1	19.7	2084.8	22.3	4.3
Chile	225.6	4.5	465.1	6.1	810.7	8.7	6.6
Peru	29.9	0.6	36.2	0.5	230.6	2.5	10.8
Mexico	28.2	0.6	53.7	0.7	102.4	1.1	6.7
Zaire	144.7	2.9	189.6	2.5	144.2	1.5	—
Zambia	402.6	8.1	580.7	7.7	607.3	6.5	2.1
Developed countries	3293.9	66.0	4669.3	61.5	4955.3	52.9	2.1
Japan	248.1	5.0	705.3	9.3	1014.3	10.8	7.3
USA	1642.6	32.9	2034.5	26.8	1682.6	18.0	0.1
Centrally-planned countries	803.9	16.1	1419.4	18.7	2322.0	24.8	5.4

Source: Metallgesellschaft AG, *Metal Statistics 1960–70 and 1970–80*, Frankfurt, 1971 and 1981.

refining. In the first decade of the 19th Century, Chile produced 15 000 tons of copper while for the decade 1961–70 copper production was 6.4 million tons. From Table 25 we see that most of Chile's production of copper concentrates are already treated within the country. There is a wide dispersion of copper mines throughout the country. While some· are very large, such as Chuquicamata and El Teniente, there are also more than 1000 small producers.

Peru produced 25 000 to 30 000 tons refined copper per year in the 1950s. The early production consisted of the complex operations of Cerro de Pasco Corporation at La Oroya where copper, zinc and lead were produced along with valuable by-products such as gold and silver. The opening of the Toquepala mine and a smelter at Ilo further increased production. In 1975 a copper refinery was commissioned to treat the blister copper produced at Ilo. In 1976 the Cuajone mine, 15 miles from Toquepala, was brought into operation, and concentrates produced there were shipped to the smelter-refinery plant at Ilo. This latter plant was expanded to take care of the increased production so that in 1980 blister copper production from Peru reached 326 000 tons. The expansion of the 150 000 ton refinery to 300 000 tons has been under discussion for a number of years.

Indonesia and Malaysia do not as yet mine sufficient tonnage to supply a minimum economic size smelting-refining complex.

The production of copper concentrates in the Philippines reached an all time high of more than 300 000 tons in 1980 — all of which was exported. However this production is scattered among a number of small and medium-sized producers who find it difficult to market their ores during periods of low copper demand. In addition, the ores of some producers contain amounts of impurities which further diminish the prices they command. With the intention of increasing the value added to the ores, the Philippine government has decided to install a 138 000 tpy copper smelter and a refinery. The total cost is estimated at $300 million, which will be largely financed by equipment credits from Japanese contractors. It is expected that 15 000 tons of the refined copper will be sold to local semi-fabricators and the remainder exported. It is important to note that the Philippine plant will produce a number of by-products such as gold, silver and sulphuric acid. Their sale will contribute to the cash flow of the operation. Important factors influencing further processing in the Philippines have been the determination of the Phillipines government that it should take place and the willingness of Japanese interests to finance the project and the existence of saleable by-products which helped to make the venture viable.

3.1.4 SEMI-FABRICATION

Most copper fabricating facilities in developing countries share certain characteristics: they are small in size, many producing a few thousands of

TABLE 25
Mine output smelted and refined in selected developing countries (Cu content, thousands of tons)

	1965[a]			1975[a]			1980[a]			
	Mine output	Smelter output	Refinery output	Mine output	Smelter output	Refinery output	Mine output	Smelter output	Refinery output	Consumption of refined copper[a]
Zambia	696	696 (100.0%)	522 (75.0%)	709	706 (99.5%)	695 (98.0%)	596	601 (100.0%)	607 (101.8%)	2.2 (0.3%)
Chile	585	557 (95.0%)	287 (49.0%)	1005	856 (85.0%)	632 (62.8%)	1068	953 (89.0%)	811 (75.9%)	42.9 (0.4%)
Zaire	289	289 (100.0%)	152 (52.6%)	444	408 (91.9%)	66 (14.9%)	460	426 (86.9%)	144 (31.3%)	3.4 (0.7%)
Peru	180	159 (88.3%)	41 (22.7%)	220	188 (85.4%)	132 (60.0%)	365	326 (89.3%)	231 (63.3%)	19.2 (5.3%)
Mexico	55	47 (85.5%)	46 (83.6%)	89	85 (95.5%)	83 (93.2%)	175	88 (50.3%)	102 (58.3%)	123.2 (20.4%)

Source: Metallgesellschaft AG, *Metal Statistics 1960–70 and 1970–80*, Frankfurt, 1971 and 1981.
[a] Percentage (in parentheses) indicates ratio between smelter or refinery output and mine output.

tons per annum, though the larger industrializing developing countries with growing electrical industries consume appreciable amounts of refined copper as can be seen, for example, from the figures for Mexico in Table 25. Other producing countries, such as Zaire and Zambia, consume very little.

Much of the world's refined copper is made into copper rods. In recent years continuous processes (continuous casting) for the production of rod (CCR) have been successfully developed. This has meant that new rod plants must adopt this technology, and that cathode copper rather than wirebar is becoming the dominant form of traded copper, as cathode copper is used directly in CCR rod production. World CCR productive capacity has increased from 200 000 tpy in 1965 to over 5 million tpy in 1980. Ninety-two per cent of this capacity is in developed countries. The average utilization rate is less than 50 per cent though there is considerable regional variation.[66] Three developing country producers, Mexico, Brazil and Indonesia, have CCR plants and their domestic markets are large enough to absorb their own rod production. Other major producers, Chile and Zambia, have entered into joint ventures with companies in the Federal Republic of Germany and France. Among factors which make it difficult for developing countries to produce and market rod are transportation problems, since rod requires careful packaging, and freight charges are high due to its high bulk-to-weight ratio. Further constraints include the need to meet customer specifications, both in regard to physical and chemical properties, and the need to resolve rapidly any customer complaints. Most movement of rod at present is over relatively short distances and it may not be easy for developing countries to develop techniques for transporting rod economically to oversea users.

3.1.5 DEMAND FOR COPPER

Developing countries' production of refined copper has traditionally exceeded their consumption by a wide margin. During the past 25 years world demand for refined copper grew at an annual rate of almost 4%; the principal uses for refined copper being electricity generation, transmission and consumption (46.3%), construction (15.9%), transportation (10.1%), industrial equipment (18.8%) and consumer appliances (8.9%).[67] It has been estimated that the rate of growth for copper consumption will decrease to around 2% per year over the period to 1990.[68] Such decline would be due to the lower average growth of real Gross Domestic Product (GDP) caused by increased energy costs and a general decline in metal consumption relative to GDP in the industrial economies.[69]

The long-term demand for copper is influenced not only by the rate of industrial expansion and by technological change but also by the price of copper relative to aluminium. The projected future 2% growth rate assumes that copper, which is less energy intensive, will be competitive with aluminium. Other production factors such as the possible introduction of electric cars, when the copper content per auto could increase from 20 kg to 100 kg[70], might make the projected increase a conservative one. Solar heating could increase the use of copper in building construction. On the other hand, optical fibre cables and microwave communications will replace copper in some telecommunications uses. Short-term demand for refined copper will, of course, be affected by the current world-wide recession.

3.2 COPPER PROCESSING TECHNOLOGY

There are two major routes to produce primary copper: pyrometallurgy, used for sulphide ores, and hydrometallurgy, used for oxide ores, and more recently, also for sulphide ores.

3.2.1 PYROMETALLURGY

Sulphide ores are concentrated, and the concentrates are subsequently smelted. There are currently six smelting processes in use: the reverberatory furnace, the electric furnace, two types of flash furnace (Outokumpu and Inco), and two types of continuous smelting (Mitsubishi and Noranda).

Reverberatory furnace smelting
The reverberatory furnace is the traditional standard of the industry and accounts even today for the greater part of copper production. In this furnace copper concentrates, either wet or roasted, are charged and smelted down to form a copper matte (a mixture of copper–iron sulphide) and a slag containing the waste minerals. The copper

matte in the molten condition is charged to a rotary furnace and air is blown through the charge using submersible tuyeres.

The oxygen of the air reacts initially with the iron sulphides to produce sulphur dioxide gas and iron oxide which reacts with added silica to form a molten slag which is removed leaving copper sulphide. Further injection of air reduces this copper sulphide to copper and sulphur dioxide. The copper so produced is impure, containing some residual iron and sulphur and needs to be refined. The refining is carried out in two stages. The major impurities are first reduced to low levels by further fire refining. Then this copper is cast into anode shapes and these are used as the positive poles of electrolytic cells containing copper sulphate solution as electrolyte. During electrolysis copper is deposited on the cathode (cathode copper). This copper is of sufficient purity to be used for casting and rolling into rod for wire production for use for electrical purposes.

This type of smelting results in the emission of large volumes of sulphur dioxide-bearing gases which are disposed of through high stacks. Present-day concern with air quality necessitates, in most cases, the installation of equipment to limit these emissions. Due to the large volumes of gases involved and the high cost of treatment, this type of furnace is no longer economic in many locations. Consequently the copper industry has been turning to other process routes.

Electric furnace smelting
This type of furnace is a type of reverberatory furnace but uses electric power as a source of energy instead of natural gas, coal or oil. It has the advantage of emitting relatively small volumes of gas with a relatively high sulphur dioxide content which can be used in acid plants. Even though electrical energy costs are high, in recent years such plants have been successfully installed in Zambia, the USA, Sweden, and Canada.

Flash furnaces (Outokumpu and INCO)
In this process, the sulphide concentrates are burnt with air or air and oxygen (Outokumpu) or with oxygen (INCO) through specially designed burners. The furnace shapes are different in the two processes and the gas strengths and volumes are also different depending upon the oxygen content of the combustion mixture.

The main advantage of flash smelting is the lower net energy requirement due to utilization of the fuel value of the iron and sulphur in the concentrates. Another advantage is the high concentration of sulphur dioxide in the gases which can readily be used for the manufacture of sulphuric acid and liquid sulphur dioxide. Outokumpu-type furnaces operate in Japan, the Republic of Korea, India, Spain, the USA, Poland, Botswana and Turkey. Other installations are planned for the Philippines, Mexico, and Chile. The INCO flash furnace process has been in use in Canada for over five years. This technology is an alternative to the Outokumpu furnace. An INCO furnace, currently under construction in the USA, will be the first such furnace used for smelting copper concentrates.

Continuous smelting (Mitsubishi and Noranda)
Processes have now been developed in which the smelting to matte and the conversion of the matte to copper are carried out in a continuous fashion. The Mitsubishi process utilizes three separate stationary furnaces, connected by launders to make material-flow continuous. The concentrates are fed through vertical lances with oxygen-enriched air and are burnt in the melting furnace. Additional heat is provided by supplementary oil burners as required. The slag and matte are separated in small electric slag cleaning furnaces and the matte overflows to a converting furnace in which an air-oxygen mixture is blown-in through vertical lances to remove any remaining iron and sulphur and to produce blister copper. There is a 60 000 tpy Mitsubishi commercial plant operating in Japan, and one is in operation in Canada.

The Noranda furnace is a cylindrical tilting furnace into which the concentrates are charged. Air or an air-oxygen mixture is blown into the furnace through submerged tuyeres and the concentrates are smelted either to a high-grade matte or to blister copper. If high-grade matte is produced this may be processed further in conventional converters to produce blister copper. These furnaces operate in Canada and process over 300 000 tons of concentrates annually. Three Noranda-process reactors using oxygen enrichment have been built at the Garfield smelter of Kennecott in the USA.

3.2.2 NEW HYDROMETALLURGICAL PROCESSES FOR CONCENTRATES

The problems that face older copper smelters with regard to air pollution are difficult and costly to resolve. The move to the newer types of processing has meant that the high-strength SO_2-bearing gases can be used for the production of sulphuric acid. This means that large quantities of acid are produced for which there may not be a ready market. Consequently, researchers have, with varying degrees of success, directed their attention to hydrometallurgical routes for which acids are needed.

Hydrometallurgical technology[71] can usually be divided into a dissolution stage which produces a copper-containing solution, a solution purification stage and a metal recovery stage. These stages are carried out at relatively low temperatures, the chemical reactions involved being dependent on the mineralogy of the copper minerals in the concentrates.

In recent years there have been several processes developed to the pilot plant or semi-commercial stage. Among these processes, the following have received some publicity: CLEAR (chloride leach); Sherritt–Cominco (sulphuric acid leach); Arbiter (ammonia leach); and Cyprus copper (chloride leach).

CLEAR process

This process has been developed by Duval Corporation, and is used in a plant of an estimated capacity of over 35 000 tons of copper per year. The process is based on the ability of aqueous solutions of certain metal chlorides to attack most metal sulphides chemically, taking the metals into solution and leaving behind a residue of elemental sulphur.

The reaction liquor is a nearly saturated chloride solution containing copper, sodium, potassium and iron. The dissolution takes place in two stages to produce a copper-bearing liquor and undissolved gangue with sulphur. After separation of the solids and liquids the solution (copper as the cuprous salt) is electrolysed in electrowinning cells. The copper is recovered as a coarse crystalline particulate metal which is then charged to melting furnaces. Further possibilities include methods for silver removal for producing a marketable sulphur product and for obtaining, by

reduction in the amount of contaminants present, a high-purity copper.

Sherritt–Cominco process

This process has been tested in a 9 tpd integrated pilot plant, and preliminary design and cost estimates have been prepared for a 75 000 ton per year plant. In the full-scale operation it is envisaged that the concentrates will be treated in two 16-hearth roasters. In the upper part of the roaster, sulphur is driven off, and in the lower reduction zone, hydrogen gas is introduced as a reductant. The sulphurous gases are used for the production of sulphuric acid while most of the reducing gases are recycled. Leaching using sulphuric acid follows during which the iron is preferentially dissolved. The H_2S evolving from the acid leach is treated for the recovery of elemental sulphur whilst the iron in solution is precipitated as jarosite (a complex basic ferric sulphate). The acid residue is then given a two stage oxidation leach in autoclaves. After solids/liquid separation, the leach solution is purified and the copper recovered as cathode copper by electrowinning.

Anaconda's Arbiter process

This process uses ammonia and involves four operations — leaching with counter current washing, solvent extraction, electrowinning and sulphate disposal. Leaching is carried out using an ammonia-oxygen mixture at near ambient pressures and moderate temperatures. Successful dissolution depends heavily upon high agitation intensities. The leached slurry is thickened and the liquors after clarification sent to a solvent extraction unit to obtain a copper sulphate solution. Ammonia is recovered from the aqueous solution by addition of lime. The copper sulphate solution becomes the electrolyte in an electrowinning circuit. A plant to produce 100 tpd of copper cathode has been in operation in the USA but is now shut down.

The Cyprus copper process

The dissolution of the copper sulphides present in the concentrates is carried out in a two-stage leach system. In the first leach, cupric ions (Cu^{++}) react with the concentrate in a strong chloride medium. After thickening and filtering the liquor, cuprous chloride is crystallized out. The mother liquor after crystallization is treated with oxygen to convert the copper and iron to the cupric and ferric states. This liquor is used to dissolve the residual copper from the first leach-separated solids (i.e. a ferric chloride leach). After thickening, the tailings are available for recovery of any molybdenite and elemental sulphur present. The crystallized cuprous chloride is reduced to metal using hydrogen as the reductant. The cuprous chloride is injected into a fluidized bed of sand, the fluidising medium being hot circulating hydrogen gas. The hydrochloric acid generated is scrubbed and recycled to the leaching circuit to maintain the chloride balance.

Despite much research and pilot plant investigations into the hydro-metallurgical treatment of copper sulphide concentrates, there has not yet appeared a process that will supersede the pyro-metallurgical route. Claims have been made by some developers that their process route can compete economically with smelting, but to date the only commercial plant operating is that of Duval Corporation, which developed the CLEAR process. The difficulties encountered in the hydro-metallurgical approach include:

(a) difficulty in solubilizing the refractory minerals such as chalcopyrite;
(b) precious metals cannot be effectively recovered;
(c) the quality of the copper produced is not as high as that produced by the electrorefining of anode copper;
(d) leach residues are difficult to handle and present a disposal problem;
(e) plant maintenance because of the aggressive nature of the solutions used.

The above comments apply to copper sulphide concentrate. If the concentrate is a mixed concentrate of two or more valuable metals, a hydro-metallurgical approach involving the roast–leach–electrowin techniques may be viable.

3.3 ECONOMICS OF COPPER PROCESSING

3.3.1 CAPITAL COSTS

The cost of a smelter/refinery complex, in 1981 terms, is likely to be over $300 million to produce 120 000 tpy copper cathode (Table 26). A typical

TABLE 26
Copper smelter and refinery capital costs (1980)

	Millions of US dollars	(%)
Materials handling	17	5.7
Smelter	58	19.3
Sulphuric acid plant	34	11.3
Copper refinery	35	11.7
General facilities	21	7.0
Construction and engineering	51	17.0
Licence fees	3	1.0
Owner's costs	32	10.7
Contingency and escalation	49	16.3
	300	100.0

breakdown of costs is given below for a flash smelter, a technology which minimizes environmental damage since it diminishes the volume of gases and increases the concentration of sulphur dioxide, permitting the economic recovery of sulphuric acid. Assuming a 10% interest rate and a 15-year cost recovery period, the annual capital charge per ton of refined copper would be US$ 329 (US 5 cents per lb).

3.3.2 OPERATING COSTS

The basic variable costs in smelter and refinery operation are for electricity and fuel. The effects of variations in these are shown in Tables 27 and 28.

When the costs of an acid plant (about $35–$50 per ton of copper) are added, and the potential by-product credits for sale of the acid ($60 per ton of copper) are subtracted, the total cost summary

TABLE 27
Annual copper smelter-refinery operating costs (120 000 tpy)

	Millions of US dollars
Labour and supervision (900 man-years at $10 000)	9.0
Fluxes: 70 000 tons ($20/ton)	1.4
Electrodes: 750 tons ($150/ton)	0.1
Operating supplies	1.9
Maintenance supplies	7.0
Indirect charges	12.2
Total	31.6

TABLE 28
Annual copper smelting and refining cost: effect
of power and electricity costs (US $ million)

Fuel oil cost (US$/ton)	150	200	310
Costs from Table 25	31.6	31.6	31.6
Fuel oil: 36 000 tons	5.4	7.2	11.2
Electricity: 180×10^6 kWh			
at 2.4 cents/kWh	4.3		
5.2 cents/kWh		9.4	
8.0 cents/kWh			14.4
Totals	41.3	48.2	57.2
Cost per ton (US$)	344.0	402.0	477.0
Cost per lb (US cents)	15.6	18.2	21.6

for copper processing in a new plant would be as
shown in Table 29, provided, of course, that the
acid can be marketed at the price assumed.

3.3.3 CONCENTRATE SALES VS. BUILDING A SMELTER

As the figures in Table 29 show, a new smelter-
refinery would need to charge at least 29 US cents
per pound ($649 per ton) to break even. Since
many existing smelters, which have already amor-
tized their capital investment or which have
access to markets for by-products, are currently
offering treatment charges which are from 5 to 10
cents per pound less than this, it is often difficult
to justify the construction of a new smelter without
government assistance (e.g. in providing low-cost
financing).

Market conditions for the sale of copper concentrates, on the other hand, make a decision to
embark on indigenous smelting very difficult. For
example, if the combined costs of processing
($649–805) with the difference between the selling

TABLE 29
Copper processing cost summary (US dollars per
ton)

Fuel/electricity costs	Low	Medium	High
Capital charge	329	329	329
Direct operation costs	344	402	477
Acid plant	36	47	59
Total	709	776	865
Less: Acid sales credits	60	60	60
Net cost per ton	649	716	805
(US cents per lb.)	(29.4)	32.5)	(36.5)

price for a ton of refined copper ($2000) and the
selling price for copper concentrates ($1300–1500)
are compared, the margin ($500–700) may not be
sufficient to cover the cost of processing.

If a flash smelter, rather than a reverberatory
furnace, is chosen there must be a market for the
resulting sulphuric acid otherwise processing costs
will increase by US $60/ton of copper (the amount
credited for acid sales).

An illustration of the constraints imposed by
environmental protection is provided in British
Columbia, which still exports most of its concentrates despite government interest in further processing. A government task force was appointed
in the early 1970s to carry out a detailed study on
the installation of a copper smelting/refining
complex. Environmental factors were found to be
of great significance. A copper smelter treating
400 000 tons of concentrates per year to produce
100 000–120 000 tons of refined copper also produces about 350 000 tpy of sulphuric acid. The
feasibility study carried out by the task force
showed, under the assumptions used, an acceptable rate of return. However, acceptable methods
to use the acid were found to be very difficult to
identify. The production of phosphatic fertilizers
and aluminium fluoride was studied but these
operations involved additional capital expendi-
tures and phosphate rock was found to be difficult
to obtain economically. The task force therefore
concluded that a case could be made for a smelting-
refining complex in British Columbia but that
there were problems which required closer study.
A decision on the installation of this plant has not
yet been made.

3.4   COPPER ROD PRODUCTION

After refining the next stage of processing is the
production of copper rod, which is being considered by some copper producers. Costs are
given below for the Southwire or Contirod continuous casting process (Table 30). Outokumpu
(Finland), Essex International (USA) and
General Electric (USA) each have smaller plants
capable of producing 1000 tons or so per month.
Capital costs of these plants are lower, although
operating costs per ton of rod are probably higher.
However, both might be viable in a limited
internal market.

TABLE 30
Copper rod plant capital costs for 120 000 tpy
plant (Southwire or Contirod)

	Thousands of US $	(%)
Melting facilities	1340	7.1
Casting machine, rolling mill	6000	31.7
Other equipment	2920	15.4
Pickling line	350	1.9
Installation	3350	17.7
Buildings	1250	6.6
Miscellaneous	300	1.6
Engineering	1250	6.6
Owner's costs	400	2.1
Contingency	1750	9.3
Total	18 910	100.0

($158 per annual ton of rod or a capital charge of $21/ton)

As there is little experience in the transportation of rod, it is difficult to assess charges for transporting rod to overseas markets. Problems of damage during transportation and of 'shelf life' must also be considered. The latter refers to the tendency of the surface of copper rod to oxidize after a few weeks' exposure to atmosphere, during transportation for example, thus resulting in a reduction in rod quality which may make it unsuitable for use for the production of magnet wire. Developing countries (Brazil, Mexico, Indonesia being exceptions) do not have a large enough home market for 120 000 tons of rod annually, most of which would therefore have to be exported. From an added value of $120 per ton in converting wire bar to copper rod capital charges of $21 and operating costs of $40–60 must be met before transport charges and import duties. Detailed studies are essential to ascertain profitability.

There is at present an import duty on copper rod in several countries among which are those countries which are major producers and users of rod. For example, the import duty into the European Economic Communities is 7.8% of the c.i.f. value which amounts to $156 per ton at a copper price of $2000.

3.5 ENERGY USE IN COPPER PROCESSING

Energy use in the copper industry varies widely among mines and processing plants. Differences in ore grade, stripping ratios, climate, altitude, type of process and age of equipment account for such variations. As ore grades decrease and stripping ratios increase, mining and concentrating become more energy-intensive. In the USA, where average ore grades have declined from 0.83% (1954) to below 0.5% at present, and waste-to-ore ratios have increased, mining and concentrating consume over 50% of all the energy used in the copper industry.[72] In the case of one mine with an ore grade of 0.7% copper and a stripping ratio of 12:1 the total energy consumed in mining was 40×10^6 BTU per ton of copper produced. In another mine where the ore grade averaged 0.55% copper and the stripping ratio was 2.5:1 the total energy consumed was 9×10^6 BTU per ton of copper produced.[73] (See Tables 31–33.)

Battelle Columbus Laboratories have studied eight open-pit mines and three smelters in the United States.[74] The smelters were green-charge reverberatory furnaces which are less energy

TABLE 31
Energy use in the production of refined copper

	10^6 BTU per ton	(%)
Mining	21.609	19.2
Concentrating	42.329	37.7
Smelting	38.222	34.0
Refining	8.185	7.2
Melting and Casting	1.947	1.7
Total	112.292	99.8

Source: Battelle Columbus Laboratories, Energy Use Patterns in Metallurgical and Non-Metallic Mineral Processing, June 1975.

TABLE 32
Distribution of energy by type (%)

Type	Mining	Concentrating	Smelting	Refining
Electrical	42.4	73.3	10.8	42.6
Gas	1.0	5.8	47.8	15.3
Oil	35.0	0.6	24.4	39.4
Coal	0.1	—	0.4	2.3
Other	21.5	20.2	16.2	0.1

Source: Battelle Columbus Laboratories, Energy Use Patterns in Metallurgical and Non-Metallic Mineral Processing, June 1975.

efficient than flash smelters. Green-charged reverberatory furnaces consume large amounts of fossil fuels and produce large amounts of sulphur dioxide which is vented to the atmosphere. This environmental problem has led to smelting processes which are less polluting.

The electric furnace is energy intensive but produces SO_2 at high enough concentrations such that production of sulphuric acid becomes practicable. The flash two-step and continuous smelting processes all use oxygen enrichment of air which reduces off-gas volumes and increases SO_2 concentration. Energy consumption is at the same time reduced.[75] In comparison to energy requirements of iron and steel and aluminium, copper presents fewer difficulties.

TABLE 33

Energy consumption per ton anode copper in smelting, converting, anode production and acid production (10^6 BTU)

Reverberatory wet charge	18.465
Electric	24.288
Flash Outokumpu	12.254
Flash INCO	9.944
Continuous Mitsubishi	13.954
Continuous Noranda	12.279

Source: H. H. Kellogg and J. M. Henderson, 1976, 'Energy Use in Sulfide Smelting of Copper'. *In:* H. K. Biswas and W. G. Davenport (editors), *Extractive Metallurgy of Copper*, Pergamon, Oxford.

4 Iron and Steel

4.1 THE PRESENT SITUATION

4.1.1 IRON ORE RESERVES

Iron ore reserves are sufficiently vast to cover world needs for at least 200 years at the present rate of extraction. Developing countries possess approximately 31% of these reserves. Quantities shown in Table 34 are based on data published by the United States Bureau of Mines (1978) as well as more recent data for the United States, Brazil, Venezuela, France and Australia.

4.1.2 IRON ORE PRODUCTION

The last quarter century has seen a tremendous surge in iron and steel production. In 1960, annual production of iron ore (measured in terms of iron

TABLE 34
World iron ore reserves

	Millions of tons	(%)
World	92 340	100.0
Developing countries	28 620	30.9
India	5580	6.0
Other Asia	1800	6.0
Liberia	630	2.2
Other Africa	1530	1.7
Mexico and Central America	360	0.4
Brazil	16 200	17.5
Venezuela	1260	1.3
Other South America	1260	1.3
Developed countries	33 120	35.8
USA	3600	3.9
Canada	10 800	11.7
France	1620	1.7
Sweden	1980	2.1
Other Europe	3240	3.5
South Africa	1080	1.1
Australia	10 620	11.5
Other Oceania	180	0.2
Centrally-planned economies	30 600	33.1
USSR	27 900	30.2
People's Republic of China	2700	2.9

Sources: United States Bureau of Mines, *Iron Ore*, Mineral Commodities Profiles, Washington, D.C., May 1978, and other sources.

content) was less than 250 million tons, while today it is nearly 500 million tons. Much of this increase in production consists of higher grade ores originating from newly exploited reserves in Australia, South Africa and South America. Brazil and Australia together accounted for 22% of the iron ore produced in 1980. In contrast, a combination of political and technical problems caused production in Liberia, Angola, Venezuela, Peru and Chile to decline in absolute terms during the 1970s.

Table 35 shows that world iron ore production grew at an annual rate of 3.6% over the last two decades and that this rate slowed to 1.5% during the last ten years. Developing countries have, on average, accounted for about 25% of total world production. Among the major producers in 1980 were Brazil (11.0%), India (4.7%), Liberia (1.9%) and Venezuela (1.8%). However, the market shares of some of these large producers have declined due to the dramatic increase in Australian production, from 3 million to 54 million tons. The rapid rise in Australian production (15.8% per annum) has contributed to a situation of worldwide oversupply and depressed prices.

Iron ore is usually sold under long-term agreements (2–15 years). The security provided by such sales has fostered a revolution in transport. Port and handling facilities were upgraded and large bulk carriers were introduced. As a result, 60% of the trade is now handled by vessels over 100 000 dwt.[76] The shift to larger vessels has reduced transport costs and allowed steel producers to rely on more distant supplies of ore.

4.1.3 AGGLOMERATION

Almost all ore is treated to increase the iron content or to obtain the desired size before the ore is used as feed in blast furnaces. Beneficiation is accomplished by crushing, grinding, screening, washing, separation, concentration and drying. Very low-grade or fine ores are often subject to agglomeration as well; either by sintering or pelletization.

TABLE 35
World production of iron ores and concentrates (iron content, in thousands of tons)

	1960		1970		1980	
	(Volume)	(%)	(Volume)	(%)	(Volume)	(%)
World	245 036	100.0	426 694	100.0	497 437	100.0
Developing countries	43 189	17.6	114 085	26.7	121 595	24.4
Brazil	3416	1.4	24 739	5.8	54 732	11.0
Chile	3625	1.5	6940	1.6	4909	1.0
India	6410	2.6	19 654	4.6	23 416	4.7
Liberia	2194	0.9	15 813	3.7	9592	1.9
Mauritania	0	0.0	5923	1.4	5530	1.1
Mexico	522	0.2	2612	0.6	3950	0.8
Peru	3139	1.3	6119	1.4	3216	0.6
Venezuela	12 668	5.2	14 080	3.3	9028	1.8
Developed market economy countries	116 496	47.6	175 244	41.1	183 943	37.0
Australia	2900	1.2	32 800	7.7	54 732	11.0
Canada	10 655	4.3	29 187	6.8	27 366	5.5
Centrally-planned economies	85 351	34.8	137 365	32.2	191 899	38.6

Source: Natural Resources and Energy Division, United Nations Secretariat, from several sources.

During sintering, the fine ore is mixed with coke breeze and fluxes and heated on a travelling grate. High temperatures partially melt the ore and bind the particles together. The screened sinter so produced tends to be friable and consequently cannot travel over long distances without producing an excessive quantity of fine material. Sintering is therefore usually carried out on the site of the blast furnace, a location that has the advantage of the availability of by-product blast furnace gas for heating purposes and coke breeze for sintering.

Another method of agglomeration is pelletizing. In this process fine-grained ore is concentrated — by gravity, magnetic separation or flotation — and compacted with moisture and a binder to form small balls (green balls) approximately 10–20 mm in diameter. The green balls are then fed to a kiln or grate and indurated by heating. After cooling and screening, the hardened pellets are ready for shipment. For successful pelletizing it is essential to grind the ore to about 0.05 mm to liberate the iron ore particles, and this grinding operation is costly. Such treatment is necessary, for example, for the taconite ores of the Mesabi Range in the USA, some magnetite ores of Sweden, the iron ore produced in Quebec, Canada and part of the Liberian and Brazilian ores.

World production of pellets has risen from practically nothing in 1955 to almost 200 million tons (iron content) in 1980, or over 40% of total iron ore production. Developing countries produce 12.5% of this amount — the largest producers being Brazil (6.5 million tons), Liberia (4.0 million), Mexico (2.5 million), Peru (2.5 million) and Chile (2.0 million). In the case of Peru, Liberia and Chile, pellets represent 35–75% of their total iron ore production. However, pelletization does not necessarily increase value, because pellet prices and demand are subject to cyclical fluctuations. During boom periods, steel-makers are willing to pay a premium for pellets (35–50%) when the sinter plant is working to capacity and the blast furnace has some spare capacity. However, during a recession, the situation is reversed. Steelmakers first buy lump ores as direct feed and then cheaper fines for the sinter plants. They buy only limited quantities of pellets. Consequently, with the current world-wide slow down in the steel industry, there is excess capacity in some pellet plants and others have been closed.

4.1.4 PIG IRON AND STEEL PRODUCTION

There are two major process routes in steelmaking: the blast furnace/basic oxygen furnace (BF/BOF) and direct reduction/electric arc furnace (DR/EAF) routes. The end product in both cases is raw steel. Intermediate products are pig iron in the BF/BOF route and sponge iron with DR/EAF.

Approximately 717 million tons of raw steel were produced in 1980 (Table 36), of which the developing countries' share was 7.2%. The five largest developing country producers were Brazil, India, the Republic of Korea, Mexico and Taiwan. Taiwan and the Republic of Korea export steel; Brazil and India are almost self-sufficient. Other producing countries such as Mexico, Venezuela and Chile must import steel to satisfy their consumption needs.

Of the developing country iron ore producers, Liberia and Mauritania currently have no home-based steel plants, while Chile, Mexico, Peru and Venezuela have been unable to satisfy internal demand for steel. Though steel-making started in Latin America more than 25 years ago, the region produces some 25 million tons of raw steel but consumes some 30 million. Most of the steel is produced by blast furnaces (84%), although direct reduction accounts for a growing percentage (16%). Among those countries with expansion plans are Brazil and Venezuela. Brazil intends to more than double existing steel-making capacity by 1985 as well as to open a new mine-steelworks complex at Carajas capable of processing 10–40 million tons of iron ore per year. Venezuela increased its steel production to 1.8 million tons in 1980 by the start-up of three Midrex series 400 direct reduction modules, making it one of the largest integrated steel producers in the world using direct reduction technology. Iron and steel

were first produced in Venezuela using conventional technology (BF/BOF), but as the country developed and steel demand expanded, further production facilities were required. Influenced by the availability of high grade ore, inexpensive natural gas as a reductant and an energy source from nearby oil fields, the adoption of the direct reduction route was justified. One original development plan envisaged the construction of a pelletizing plant, three HyL units, three Midrex units, two electric furnaces, two continuous casting plants, a continuous wire mill and extensions to the rolling mill facilities. Chile produces almost 500 000 tons of steel per year using the blast furnace/basic oxygen furnace route. Semi-processing involves the use of rolling mill, tube and pipe mills, electrolytic tinning and galvanizing lines. Peru has a small steel operation with a pig iron capacity of around 250 000 tons, and small semi-processing plants produce rolled steel, shapes, rods, bars, etc. as well as some tin plate and galvanized sheets. Plans have been made to expand production to about 750 000 tons of steel and the direct reduction/electric furnace process route will be used.

India produced 9.5 million tons of steel in 1980, which matched steel consumption. Planned expansions are in some difficulty because of a shortage of electrical power.

Integrated steelworks have been operating in developing countries of Africa and the Middle East since the late 1950s. Combined production was 3 million tons, compared to apparent con-

TABLE 36
World raw steel production

	1970		1980	
	(Millions of tons)	(%)	(Millions of tons)	(%)
World	596.3	100.0	717.2	100.0
Developing countries	25.5	4.3	51.7	7.2
Brazil	5.4	0.9	15.3	2.1
India	6.3	1.1	7.1	1.0
Republic of Korea	0.5	0.1	8.6	1.2
Mexico	3.9	0.7	7.1	1.0
Taiwan	0.6	0.1	4.2	0.6
Developed market economy countries	396.9	66.5	409.8	57.1
Centrally-planned economy countries	174.0	29.2	255.7	35.7

Source: International Iron and Steel Institute

sumption of 21 million tons in 1978. Algeria, Egypt and Tunisia use local iron ore supplies in meeting their internal demand for steel. Algeria is considering proposals to build a major steel works, and discussions have been held with Japanese, Canadian and American companies as well as the Soviet Union. A number of other energy-rich Arab countries would like to increase Arab steel-making capacity from 3.5 million tons to 20 million tons.[77] Saudi Arabia, Egypt, Libya and Morocco all have projects planned. Of these countries, only Libya and Algeria have sizable iron ore reserves.

It would appear that co-operation among the iron ore-rich African countries (Liberia and Mauritania) and energy-rich Arab countries offers an opportunity to realize gains from further processing.

4.1.5 DEMAND FOR STEEL

Internal demand in developing countries
In 1980 developing countries produced 52 million tons of raw steel and consumed 61 million tons. At present 50 developing countries possess steel-making capacity, but it is not expected to keep pace with internal demand. Newly industrializing countries and oil-rich countries in particular will experience a continuing demand for more steel.

Another indication of the extent of possible future growth of internal demand is the disparity between developed and developing country *per capita* consumption (Table 38).

World demand
Developing countries produced 24% of the world's

TABLE 37
Steel capacity, production and consumption in developing countries[a] (millions of tons)

	1977	1985	1990
Capacity	100.0	180.0	250.0
Capacity utilization rate (%)	76	77	79
Production	75.9	138.6	195.5
Consumption	101.2	173.0	253.0

Source: Nijhawan, B. R., *Global Scenario of World Steel Industry Growth, Particularly up to 1985,* Steel in the 1980's, Paris Symposium, February 1980, OECD.
[a] Includes centrally-planned Asia, China, the Democratic Republic of Korea and Southern Europe so that totals are larger than those used elsewhere in this chapter.

TABLE 38
Steel consumption *per capita*[78] (1978)

Region	Steel consumption per capita (kg)
North America	662
Western Europe	320
Oceania	273
Middle East	99
Latin America	95
Far East	59
Africa	34

iron ore and 7.2% of its steel in 1980.

The demand for iron ore grew at 3.6% annually during the 1960–80 period, but is expected to grow at a lower rate (2.4–2.8%) to 1990.[79] Given a situation of excess supply and a decline in demand, the prices of iron ore declined in nominal and real terms during the 1976–80 period. Historically, the real price (c.i.f.) of iron ore has declined because of the large increase in supply and the improvements in transport and handling. In 1979 and 1980, however, transport costs rose dramatically, increasing the c.i.f. price, but the f.o.b. price did not rise and even declined in real terms. In general iron ore prices lag behind inflation, due to long-term contracts.

About a third of traded iron ore moves under long-term contracts, one third moves on the spot market, and one third via intrafirm transactions (captive mines). Continued inflation has caused a shift from inflexible long-term contracts to ones with adjustable 'recession'-type clauses. Terms have already been adjusted between Japanese importers and exporters in Australia, Peru and Brazil.[80] The last peak year for steel was 1974. Since the demand for steel is directly related to the rate of economic growth, lower GDP growth rates due to the energy crisis and a levelling off in the metal-intensiveness of advanced industrial economies is resulting in continued lower growth in the demand for steel. Consequently, the annual rate of growth in steel demand is expected to decline from 4.1% (1960–80) to less than 3% in 1980–90. However, this shift masks the variation in the rates of growth among country groups. Demand in industrialized market economies (which consume 50% of world steel production) is expected to grow at only about 1–1.5% annually, as compared to 4–6% for developing countries.[81]

Because the industrialized countries continued to expand capacity at a time when demand was falling they are presently suffering from excess capacity and this situation is expected to continue through 1985. Such overcapacity in industrialized countries is fuelling the trend towards protectionism. This may limit developing countries' steel exports. Given the fact that developing countries have not and will not be able to satisfy their own internal demand however, there is room for expansion of processing on a country or regional basis.

4.2 IRON AND STEEL TECHNOLOGY

Much technical progress has been made in the iron and steel industry in the developed countries, where it is one of the major industries, employing hundreds of thousands of workers, both directly and indirectly, and is an important earner of foreign exchange. The developed countries rely, in the main, on importing large quantities of foreign ores. These ores are reduced to pig iron in blast furnaces, blown to steel in oxygen converters and cast continuously into slabs for further processing. Such techniques are not suitable for relatively small-scale production, for which other routes have to be considered.

4.2.1 DIRECT REDUCTION STEEL MAKING

There are many patented processes for the direct reduction of iron ores, but very few have been developed commercially. One such successful operation is the 'HyL' process. A 200 ton per day plant was installed at the Hojalata y Lamina S.A. plant at Monterrey, Mexico in 1957 and since that date plants have been built in Brazil, Venezuela, Indonesia, Iran, Iraq and other countries.

In the HyL process, iron ore is reduced to sponge iron with a mixture of carbon monoxide and hydrogen gas. The reaction takes place in a fixed-bed reactor, where the gas flows downward through lump ore or pellets. The reducing gas is generated by steam reforming of natural gas or other hydrocarbons. The process operates through a four-step cycle, and the four reactors used are programmed so that each of the stages of the process is always underway in one of the reactors.

The other process that is achieving widespread usage is the Midrex process developed by the Midland–Ross Corporation and sold to Korf Industrie of the Federal Republic of Germany.

In this process, a synthetic gas, composed of hydrogen and carbon monoxide, is used as reductant. The reactor is a shaft-type furnace into which iron oxide pellets are charged continuously at the top and from which sponge iron is withdrawn continuously from the bottom.

These two processes together account for more than 60% of all world production of reduced iron, which is over 15 million tons per year.

Processes using coal as a reductant have also been developed, in particular by use of the Stelco–Lurgi (SL/RN) process. This process uses a rotary kiln for the reduction stage. The ore is charged to the furnace and the reductant may be coal, coke breeze or lignite. The solid products from the kiln are discharged through a gas-tight chute into a cooler. After cooling, the products are screened and the residual char recycled to the kiln.

One such application is the plant of New Zealand Steel Ltd. where a SL/RN kiln was commissioned in early 1970. After major process changes, particularly in the direct reduction operation, a process route for the production of billet steel has been developed using indigenous iron sand and a non-coking highly reactive coal.

The ACCAR process is similar to the SL/RN but is still at the test stage. An ACCAR plant to produce 150 000 tpa sponge costing $25 million started operation in Orissa, India in 1981. This plant uses domestic lump ore and locally available coal as reductant.

Of the many direct reduction processes, the three described above have achieved some commercial success. The processes using natural gas as a reductant have become the leaders in the field, and there is today adequate knowledge of the costs involved to assess accurately their respective merits. What has become clear in recent years is that a steel industry of 500 000 to 1 000 000 tons per year can be a viable operation in a developing country.

4.2.2 CONVENTIONAL STEEL PRODUCTION

The conventional process for steel production is the treatment of pig iron (produced in a blast furnace) in basic converter furnaces (B.O.F.). The use of oxygen converters has virtually superseded open hearth furnaces due to lower costs.

The direct reduction/electric arc furnace (DR/EAF) route is now also well established, but whilst viable at low production (say modules of 500 000 tons) it cannot compete economically with the blast furnace/basic oxygen furnace (BF/BOF) for large scale production. At lower production rates it is fully competitive especially when cheap natural gas is available.

4.3 STEEL PROCESSING COSTS

4.3.1 CAPITAL COSTS

Table 39 shows the 1980 costs of a plant which produces 500 000 tons per year of steel based on the direct reduction (DR) of iron ore followed by electric furnace melting (EAF) and continuous casting of steel into slabs. There is at present no international market for metallized iron ore, due to its propensity to oxidise during transit. Metallized iron ore is normally melted as soon as practicable after production in electric furnaces to

TABLE 39
Capital costs of DR/EAF plant

	Millions of US dollars
Site development	48
Storage buildings	4
Direct reduction plant	75
Steelmaking plant (electric furnace)	53
Continuous casting plant	50
Miscellaneous	10
	$240

produce steel for casting into slabs.

The capital cost per annual ton is $480, equivalent to an annual capital charge of $63.

4.3.2 OPERATING COSTS

Table 40 shows estimated operating costs (in 1980 US dollars) for a 400 000 ton capacity sponge iron plant.

The effect of natural gas and electricity prices on total costs is indicated in Table 41.

TABLE 40
Sponge iron operating costs excluding energy

	Millions of US dollars
Iron ore: 1.5 tons per ton product, at $35/ton	21.0
Labour and supervision: 60 man-years at $10 000	0.6
Maintenance supplies $4 per ton product	1.6
Miscellaneous	0.7
Total	$23.9

4.3.3 REVENUE

The selling price of slabs is of the order of $280 per ton. The combined capital charge ($63) and operating costs ($191–323) give a cost range of $254 to $386. At the lower range of costs the operations can be economically viable. The effect of energy costs gives those projects with sources of low-cost energy a significant advantage.

TABLE 41
Effect of gas and electricity prices on operating costs

	Natural gas prices (US $/million BTU)		
	4	8	12
Basic operating charges (Table 40)	23 900 000	23 900 000	23 900 000
Natural gas (12 500 000 BTU per ton product)	20 000 000	40 000 000	60 000 000
Electricity (150 kWh per ton):			
2.4¢/kWh	1 440 000		
5.2¢/kWh		3 120 000	
8.0¢/kWh			4 800 000
Total	45 340 000	67 020 000	88 700 000
Cost per ton	$113	$167	$222

4.3.4 ECONOMIC FACTORS

Conditions favourable to DR/EAF and continuous casting operations include cheap iron ore, low energy costs, natural gas as a reductant and a sufficiently large market to act as a base-load outlet for the product.

Costs for production of steel in a 500 000 ton per year electric arc facility are shown in Table 42.

4.4 ENERGY USE IN IRON AND STEEL

The energy required to produce ores, concentrates and agglomerates is summarized in Table 43. The data in Table 43 are based on a survey of the United States steel industry in 1973. There were three major process routes and their respective shares of production were: BOF (55.2%), open-hearth (26.4%) and electric arc furnace (18.4%).

All three processes produce liquid steel which can be transformed into mill products using either

TABLE 43
Energy use in mining and agglomeration (10^3 BTU per ton of agglomerates)

Mining	606		
Crushing	134		
Concentration	1011		
Pelletization	561	Sintering	2470
Miscellaneous	309		
Total	2620		4530

Source: Battelle Columbus Laboratories, *Potential for Energy Conservation in the Steel Industry*, May 1975.

TABLE 42
Electric arc steel operating costs (melting and casting) at an annual capacity of 500 000 tons steel

Operation	Annual charge ($)		
Smelting and casting			
Scrap iron 150 000 tons at $80 p.t.	12 000 000		
Labour and supervision: 300 at $10 000 p.a.	3 000 000		
Fluxes: 50 000 tons at $35 p.t.	1 750 000		
Ferroalloys: 10 000 at $600 p.t.	6 000 000		
Refractories: 20 000 tons at $400 p.t.	8 000 000		
Electrodes: 3000 tons at $2000 p.t.	6 000 000		
Maintenance and supplies	2 400 000		
Miscellaneous	1 050 000		
Total	40 200 000		

	Natural gas prices ($/million BTU)		
	4	8	12
Smelting and casting	40 200 200	40 200 200	40 200 000
Natural gas at 400 000 BTU p.a.	1 600 000	3 200 000	4 800 000
	41 800 000	43 400 000	45 000 000
Electricity 700 kWh p.t.			
at 2.4¢ per kWh	8 400 000		
at 5.2¢ per kWh		18 200 000	
at 8.0¢ per kWh			28 000 000
Totals	50 200 000	61 600 000	73 000 000
per ton	100	123	146
Combined sponge iron and slab production			
Annual sponge iron costs Range: ($ million)	45.34 − 67.02 − 88.70		
Annual melting and casting costs Range: ($ million)	50.20 − 61.60 − 73.00		
Total	95.54 −128.62 −161.70		
Final operating cost per ton slab	$191 $257 $323		

TABLE 44
Energy use in steelmaking (10^6 BTUs per ton)

Blast furnace	Open hearth	Ingot casting	Slabbing mill	Total for steel slabs
17.37	6.12	0.741	1.91	26.15
Blast furnace 21.81	Basic oxygen 2.77	0.741	1.91	27.24
Electric arc 10.03	—	0.741	1.91	12.69
Blast furnace 14.932	Open hearth 5.26	Cont. casting 1.07		21.26
Blast furnace 18.74	Basic oxygen 2.38	1.07		22.20
Electric arc 8.62		1.07		9.70

Source: Battelle Columbus Laboratories, *Energy Use Patterns in Metallurgical and Non-Metallic Mineral Processing*, June 1975.

a non-integrated or a continuous casting process. The non-integrated process would consist of casting ingots, soaking them in heated pits and putting them through a slabbing or blooming mill. Continuous casting by-passes the ingot stage and transforms liquid steel into billets, blooms and slabs. Continuous casting saves energy since it eliminates heat lost in ingot casting and the soaking pit. Yields are also higher so that more products are made from the same amount of liquid steel.

Energy consumption is influenced by the process route, by the degree of integration of the various stages, by the use of scrap, by plant utilization and by the age of the equipment. Integrated operations allow the capture and use of by-product energy such as BF/BOF gas. Low rates of utilization increase energy consumption since the base-load energy consumption is independent of the level of operating activity. One hundred per cent use of scrap in the electric arc furnace results in the lowest energy consumption. The blast furnace can absorb about 30% scrap but is a poor melter of scrap.

United States steel plants use about the same amount of energy per ton as United Kingdom plants, 13% more than West German plants and 24% more than Japanese plants. The German and Japanese plants are newer, use the basic oxygen furnace and are more highly integrated (Table 45).

The American Iron and Steel Institute produces annual reports on the energy consumption of its members. In the past six years the USA has shifted to the basic oxygen furnace (61% of production) and electric arc furnace (25%) and away from the open hearth (14%). In 1978, 29.9 million BTUs were consumed per ton of steel shipped. In its annual statistical report of 1979, the American Iron and Steel Institute published data on energy consumption as broken down among the following sources: coal 63.5%; gas

TABLE 45
Western steel industry energy use in selected countries

	USA	UK	FRG	Japan
Basic oxygen furnace (per cent of total production)	56	48	70	80
Share of steel continuously cast (per cent)	6	3	17	21
Million BTU per ton raw steel	25	25	22	19
Million BTU per ton steel shipped	34	35	29	26

Source: Battelle Columbus Laboratories, *Final Report on Potential for Energy Conservation in the Steel Industry*, Columbus, May 1975, 11134–38.

20.6%; oil 10.5%; and electricity 5.3%.

In most developing countries the direct reduction/electric arc furnace route offers the best possibility to expand processing since it is not as capital-intensive as the BF/BOF route. Direct reduction has been used for over 100 years. It produces iron in a solid state (sponge iron) and permits the use of fuel other than coking coal, such as natural gas. A study of direct reduction plants in Bolivia, Venezuela, Peru and Iran revealed that the energy required to produce 90% metallized sponge iron in 1000 ton-per-day plants ranged from 12 to 23 million BTU per ton of iron. Under the most favourable conditions, the minimum total energy required by this route is 18 million BTU per ton of continuously cast slabs.

A comparison of the relative energy consumption of the different routes is given in Table 46.

Energy consumption in steel making varies by process route and by use of scrap (see Table 47).

Those countries with large iron and coal production may use the BF/BOF route; those with smaller deposits (or markets) and oil, gas or hydropotential may choose DR/EAF. Though much has been written about the higher energy-intensity of the DR/EAF route (26.7 million BTU per ton liquid steel) it is actually less energy intensive than a BF/BOF route which uses no scrap (27.5 million BTU per ton liquid steel).[82]

TABLE 46

Relative energy consumption (10^6 BTU per ton slabs)

BF/BOF/ingots/soaking pits/slabbing mill	20.6
BF/BOF/continuous casting	16.8
Electric arc furnace (100% scrap)/continuous casting	7.6
DR/EAF (25% scrap)/continuous casting	22.0

Source: Battelle, *Final Report*, 1975, p. V 60, Data are from a 1971 study.

TABLE 47

Effect of operating strategies on energy requirements

Operating strategy	% scrap	Energy required[a] (10^6 BTU/ton liquid steel)
Blast furnace + BOF w/ore trim	0	27.5
Electric furance + 100% direct reduction ore	0	26.7
Blast furnace + BOF w/scrap + ore trim	25	22.6
Blast furnace + BOF w/scrap trim	30	21.7
Electric furnace + 30% direct reduction ore	50	16.9
Blast furnace + open hearth furance w/O_2	50	15.8
Blast furnace + open hearth furnace w/O_2	70	12.5
Hot blast cupola + BOF w/scrap trim	100	8.6
Electric furnace	100	7.0
Open hearth furnace w/O_2	100	6.7

Source: Field, Lewis, "The Impact of Energy Conservation", *Iron and Steel Magazine*, January 1979, p. 10.
[a] Scrap energy value = 0

5 Economic factors affecting the location of processing in developing countries

5.1 CAPITAL REQUIREMENTS

Intensity of capital use may be measured by the capital–labour ratio, which also serves as a measure of the degree of mechanization as well as of labour intensity.[83] The capital–labour ratio is the ratio of the stock of investment in fixed capital and in working capital to the flow of labour working with it. In practice, working capital is ignored and since the actual value of the stock of investment is difficult to estimate the capital stock is measured by accumulating investment over a number of years.[84]

By combining data from a recent Economic Commission for Europe survey of European industry and the annual United States survey of manufacturing, it is possible to rank various industries according to their capital intensity (Table 48).

Capital intensity in Europe was measured by summing gross domestic capital formation from 1959 to 1970 and dividing it by the average of total employment in 1958–60 and 1968–70. United States capital intensity was obtained by dividing the capital stock by the number of employees in 1968.

From Table 48, it is seen that the basic metal industry is the most capital-intensive after the chemical industry. Low on the list is metal-fabrication which ranked 6 out of 8 without exception. Developing countries may question the relevance of European and United States rankings but capital requirements are a function of technology and product. Since in heavy industry, particularly basic metals, the range of commercially feasible technologies is narrow, developing countries must use the technologies in use in the developed countries.[85]

In a recent UNCTAD–UNIDO modelling project studying the impact of technology on capital requirements, countries were classified according to degree of industrialization and various technological variables were calculated for different industries. Capital intensity was estimated for iron and steel, non-ferrous industries, metal products, food, textiles and wood. The results are indicated in Table 49.

Basic metals are clearly more capital-intensive than any other type of manufacturing. If estimates of minimum efficient size are combined with those for current capital cost per ton and labour

TABLE 48
Capital intensity of various industries (1 = most capital intensive, 8 = least capital intensive)

	Belgium	Finland	France	FRG	Italy	Netherlands	Norway	Sweden	UK	USA
Chemicals	1	2	1	1	1	2	2	2	1	1
Basic metals	2	1	3	2	2	1	1	1	2	2
Non-metallic minerals	3	5	2	3	4	3	3	4	4	3
Food, drink, tobacco	5	4	4	4	5	4	4	5	3	5
Paper, printing	4	3	5	5	3	5	5	3	5	4
Metal-fabrication	6	6	6	6	6	6	6	6	6	6
Textile, clothing, footwear, leather	8	8	7	7	7	7	8	8	7	7
Wood	7	7	8	8	8	8	7	7	8	8

Sources: Economic Commission for Europe, *Structure and Change in European Industry*, New York, United Nations, 1977; United States Department of Commerce, *Annual Survey of Manufactures: Industry Profiles*, Washington, D.C., 1972.

TABLE 49
Average capital/labour ratio 1967–76 (1970 US dollars)

	Large highly industrialized[a]	Small highly industrialized[b]	Small semi-industrialized[c]	Large semi-industrialized[d]	Low industrialization[e]
Iron and steel	$42 897	$27 535	$12 869	$10 267	$8678
Non-ferrous	$39 927	$39 716	$16 015	$10 453	$8539
Food	$27 075	$20 137	$9207	$4776	$6828
Metal products	$18 090	$9962	$4099	$4496	$2830
Textiles	$17 223	$13 617	$5391	$4193	$4042
Wood	$17 176	$13 923	$6362	$4733	$2914

Source: Joint UNIDO–UNCTAD Global Modelling Project, Geneva, 1980.
Notes: *Average Capital* represents the sum of gross investment over ten years modified by average life of plants and equipment and the rate of growth of capital accumulation.
[a] USA
[b] Six countries: Australia, Canada, Denmark, Finland, Norway, Sweden.
[c] Ten countries: Colombia, Egypt, Greece, Iraq, Israel, Peru, Philippines, Portugal, Singapore, Venezuela.
[d] Four countries: Brazil, Korea, Spain, Turkey.
[e] Ten countries: Bolivia, Cyprus, Ecuador, Ethiopia, Guatemala, Hong Kong, Mozambique, Nigeria, Panama, Tunisia.

productivity, total capital and labour require-ments as well as specific capital intensities (capital per job) can be derived (Table 50).

Analysis of capital, labour and materials and energy costs help to indicate in specific cases whether a cost advantage may be made use of by the potential host country to attract investment. Average costs recalculated as percentages have been calculated for aluminium, copper and steel and are set out in Table 51.

The following energy costs were used: alumina, $200/ton oil; aluminium 2.4 cents kWh and $150/ton oil, 5.2 cents kWh and $200/ton oil; copper $200/ton oil; iron and steel, $4 per 10^6 BTUs gas and 2.4 cents kWh; $8 per 10^6 BTUs gas and 5.2 cents kWh.

The share of capital in total annual costs is larger for aluminium and steel than for copper.

The annual capital charge is a function of the interest rate and payback period. Capital costs are quite large for all three minerals, so that sources and terms of financing are critical.

5.2 SOURCES OF FINANCE

Capital costs of mining and processing complexes have escalated during the last decade. In 1980 the Chairman of Rio Tinto Zinc noted that the capital costs per ton of annual capacity for a combined mine-processing operation have risen from $2000 to $8000 per ton of refined copper and from $1500 to $4000 per ton of aluminium.[86]

Rising capital costs have changed the manner in which mineral development is financed. Pre-viously, mining corporations were able to finance new expansions from equity, from retained earn-

TABLE 50
Capital and labour requirements and capital/labour ratio

Commodity	Minimum economic size (tpy)	Capital costs per ton (1979)	Total capital cost (1979)	Output per man year (tons)	Total man years	Capital per job (1979)
Alumina[a]	500 000	$700	$350 000 000	800	625	$560 000
Aluminium[a]	100 000	$4000	$400 000 000	90	1100	$360 000
Refined copper[a] (smelter and refinery)	120 000	$2500	$300 000 000	140	860	$350 000
Steel DR/EAF[a]	600 000	$400	$240 000 000	200	3000	$80 000
BF/BOF[b]	1 000 000	$820	$820 000 000	200	5000	$164 000

Sources: [a] Estimates for 1979; [b] UNIDO, *Mineral Processing in Developing Countries*, p. 82.

TABLE 51
Estimated share of capital, labour, raw materials and energy in total annual costs per ton (%)

Commodity	Capital	Labour	Principal raw material	Complementary raw materials	Energy	Miscellaneous
Alumina	34	6	23	5	26	6
Aluminium						
Low cost energy	33	6	30	7	22	2
High cost	26	5	24	6	38	1
Refined copper	15	3	65	2	7	8
Steel (DR/EAF)						
Low cost	25	3	17	26	25	4
High cost	20	2	13	22	40	3

ings or from commercial bank loans. However, such sources are clearly inadequate when base-metals projects cost in excess of one billion dollars. Another new factor is the increasing debt-to-equity ratios of transnational mining corporations. It is frequently stated that until the 1960s, 90% of the capital costs were financed from equity or retained earnings of the mining houses.[87] Bank loans have always played a role in mine development; commercial loans were instrumental in consolidating the diamond mines in South Africa as well as in opening copper mines in Peru and Zaire. But it is no longer possible for a single mining corporation to open a major new mine relying solely on internal cash flow and bank loans.

Given the size of projects, it has become necessary to use a number of novel mechanisms involving many lenders; however, there are a number of other reasons, in addition to large capital requirements, that explain the increase in the number of participants in project financing.[88] The growing concern over security of supply of metals has caused the German and Japanese governments to become involved in mine development. There is also the desire on the part of transnational corporations to minimize non-commercial risks by inviting the participation of international organizations. The twin objectives of security of supply and minimization of risks have caused a number of developed countries to offer, on an individual basis, investment insurance to the transnational corporations. However, such insurance has proved to be inadequate to the need and as a result the European Economic Community is considering bilateral contracts between the community and the investor to guarantee against wars, expropriations and other unilateral modifications which would have an adverse effect on the viability of the investment. The firms wishing to have the guarantee would be asked to pay a premium. The receipts from the premiums would be intended to ensure the financial autonomy of the guarantee mechanism.[89]

Large mining complexes are now often financed by project loans extended by a consortium of lenders. Such loans are often accompanied by bilateral agreements, by loan guarantees and by waivers of sovereign immunity. To facilitate the financing of the Cuajone mine-smelter complex the Peruvian government and the Southern Peruvian Copper Corporation (SPCC) signed an agreement stipulating benefits, guarantees, a taxation system and foreign exchange rules to be in force during the recovery of investment. Such a bilateral agreement enabled the SPCC to obtain project loans from eleven different groups including suppliers of equipment, copper buyers, the US Export–Import Bank, the International Finance Corporation and 29 commercial banks.[90]

The terms of such loans can be demanding and are, of course, designed to insure that the lenders are repaid first. For example, escrow accounts are often set up outside the producing country to receive sales revenues, allowing banks to control disbursements including loan repayments. Deficiency agreements are also included which make the project sponsors directly liable for any debt service not met by the project itself. The payback periods are medium-term, generally between 5–7 years, and interest rates are periodically adjusted to maintain a spread between the bank's own cost of borrowing and the rate paid by the borrower. Lenders may also require that the project be managed and operated by a mining company which has the necessary management and technical skills and experience in marketing.

Financing arrangements for smaller ventures are less complex but contain the similar mechanisms to those used in the Cuajone scheme. The recently announced Philippine copper smelter-refinery will be financed, in part, by Japanese supplier equipment credits. While sufficient in this instance, this method has a number of drawbacks. Equipment credits cover only a portion of total expenses and are of little use to developing countries which possess mining equipment industries.

Consumer credits have been used by the Japanese to start a number of Canadian mines. The long-term nature of the credits is a distinct advantage. Production credits are usually given by banks against the minerals in the ground and are recovered from sales revenues. This method of financing requires stable prices and the host countries' sovereignty will be infringed if banks require a legal claim on minerals as security.[91]

Much has been written about the new sources of development capital such as oil companies, oil-exporting countries, insurance companies and merchant financiers. While any of these entities are possible participants in consortium lending it is unlikely that they would finance mineral projects alone. Shell, for example, joined with others in financing Cuajone, and Amoco is involved in a consortium to develop the Ok Tedi Mine in Papua New Guinea. Insurance companies are, in general, risk averters and prefer security and guarantees for their loans.

It seems doubtful that resource-rich developing countries and oil exporting countries will finance billion dollar mining projects without the participation of mining companies or international organizations.[92] It seems more probable that available funds of the oil exporting countries will be channelled through the World Bank or commercial banks as in the past. The consortium approach appears to best meet the needs of both lenders and borrowers. Thus, for the immediate future, mineral development seems likely to occur with the active participation of mining companies, international financing organizations, and producing countries.

5.3 LABOUR REQUIREMENTS, LABOUR INTENSITY, LABOUR SHARE IN TOTAL COSTS

Labour intensity in mineral processing is low, compared to other branches of manufacturing, particularly in comparison with textiles, clothing, footwear, leather and wood products. Within the range of mineral production, mining and processing are less labour-intensive than some types of semi-fabricating. Indeed the labour intensiveness of semi-fabricating is on a par with that of textiles and wood according to UNIDO–UNCTAD data. The share of labour costs in processing is too small (2–6%) to give developing countries a significant competitive advantage even though wages vary considerably (Table 52).

To examine labour requirements purely in terms of labour intensity or of its share in total costs neglects the question of skills. The import-

TABLE 52
Annual wages and salaries per employee (average for 1972–76) 1970 US dollars

Industry	Country					
	Large highly industrialized	Small highly industrialized	Small semi-industrialized	Large semi-industrialized	Low industrialization	India
Iron and steel	$10 561	$4632	$1909	$2212	$1415	$657
Non-ferrous	$9212	$5402	$1971	$1886	$1141	$712
Metal products	$8092	$4088	$1388	$1413	$855	$444

Source: UNIDO–UNCTAD Modelling Project, Geneva 1980 (country groups as shown in Table 49).

ance of technical and managerial skills is demonstrated by efficient production and high quality output. Most discussions of mining and mineral processing emphasize their capital intensiveness without mentioning the fact that capital-intensive operations are usually also skill-intensive.[93]

Host countries need to know what skills are required at each stage of processing and the length of time needed for the formation of such skills. In Zambia for example, 15 years after independence, efficient mining operations are still limited by the availability of mining engineers.[94]

A minimum size alumina plant employs between 700–800 workers; an aluminium smelter needs 1000–1600. In both cases half of the labour force must be semi-skilled, one-third skilled; one-tenth must have varying administrative/clerical skills and the remaining tenth scientific and technical training. Trained personnel, such as production managers and superintendents, process engineers and technical foremen, are essential if costly interruptions in potline operations are to be avoided. It takes three to four years to train a group of potroom operators.[95]

The EEC undertook a study of the structure and changes in European industry from 1958 to 1970 in which it examined labour skills by branch of industry. It found that in eight out of nine countries studied the skill requirements in basic metals processing exceeded those for metal products. It also found that more scientists and engineers were employed in petroleum and coal, electrical machinery, chemicals and basic metals than in other branches of industry.

5.4 RAW MATERIALS REQUIREMENTS

The share of the cost of raw materials (ores, concentrates, complementary inputs) in the cost of refined metal is variable. The highest share is for copper (66%) and raw steel (43%) and lowest for alumina (28%) and aluminium (30–37%). Important savings can be realized by increasing the efficiency with which inputs are used, particularly by decreasing their handling and transport costs.

The most important material inputs for alumina are bauxite and caustic soda; for aluminium, they are alumina and electrodes. Alumina is also land-intensive; 100 hectares (ha) are needed for buildings and the red-mud ponds. To produce one ton of alumina 2.2 tons of bauxite, caustic soda (0.095 tons) and fuel oil (0.35 tons) are needed. One ton of aluminium requires 1.93 tons of alumina, 0.6 tons of electrodes, 0.06 tons of electrolytes and 14 000 kWh of electricity.

Since alumina is material-intensive, refineries tend to be located near bauxite deposits or at least near a port. The alumina and energy-intensiveness of aluminium on the other hand favour the location of aluminium production near a source of cheap energy, especially since the fall in the transport costs of bauxite and alumina.

The importance of copper ores and concentrates in total costs and the tremendous reduction in bulk/weight during processing explains, in part, why developing countries smelt and refine so much of their mine production. Since concentrates are three-quarters waste, considerable savings are achieved by smelting near the mine. However, since the smelted product is 95% copper there is no saving in transport costs from refining near the mine.

In some cases the need for complementary inputs will determine the location of plant. In the case of iron and steel, for example, according to a UNIDO study, in Mexico, one ton of finished product requires 1.6 tons of iron ore, 1.1 tons of coal, and 0.4 tons of limestone; this combination of inputs yields one ton of pig iron which, when combined with 0.7 tons of scrap iron, yields 1.4 tons of ingot steel which in turn yields a ton of finished products.

The availability of suitable raw materials is a major factor in the cost of steel production. A lack of good coking coal explains the high cost of Brazilian steel as compared to Mexico and India. In the direct reduction-electric arc furnace process the availability of high quality ore or pellets and scrap is essential.

5.5 TRANSPORT COSTS

Producing countries which are close to major consuming markets have a competitive advantage over more distant producers, since their delivered costs are lower. Where a country's production of processed metals exceeds its needs, transport costs may determine its ability to sell in international markets.

TABLE 53
Size of ships transporting iron ore and aluminium raw
materials (percentage share of seaborne trade in mineral)

Size of ship	1970	1972	1974	1976	1978
25 000 dwt iron ore	24	15	12	9	8
bauxite/alumina	21	48	43	43	—
25–40 000 dwt iron ore	19	15	10	8	5
bauxite/alumina	32	38	36	26	—
40–80 000 iron ore	49	46	39	32	27
40 000 dwt bauxite/alumina	7	13	20	31	—
80 000 dwt iron ore	11	24	39	51	60

Source: Fearnley and Egger's World Bulk Trades.

Iron ore and bauxite are among the five most important commodities in dry bulk transport, accounting for almost 50% of total tonnage shipped in 1978. There has been a revolution in recent years in dry bulk transport and long-term forward contract sales have facilitated investment in large bulk carriers. The following two tables illustrate this evolution in vessel size and its impact on costs.

The most common vessel used in the transport of iron ore is in excess of 100 000 dwt; for bauxite it is 25 000–60 000 dwt and for alumina up to 25 000 dwt. In contrast, copper cathodes and wirebars and all semi-fabricated products are not usually carried by dry bulk carriers but by freight liners.

Almost all aluminium raw materials are trans-ported under long-term arrangements made by the six transnational corporations. In most cases the aluminium companies enter into contracts of affreightment or cargo guarantees with the shipping companies, but in North America, company-owned vessels supplemented by time charters are used. For example, Alcoa Steamship Co. controls a number of dry bulk carriers some of which are co-owned, others are time and voyage chartered.

Japan and the EEC steelmakers enter into long-term contractual purchases and shipping arrangements with national shipping contracts usually using contracts of affreightment and cargo guarantees of 8–15 years.[96] US steel companies use their own or associated ore carriers. Of the iron ore producers, only India and Brazil operate ore carriers, but Venezuela and Peru have stated

TABLE 54
Comparative ocean transport costs of different size vessels

Ore	1970 iron ore[a] (US $ per ton)		1976 bauxite[b] (US $ per ton)		1978 dry bulk[c] ($/dwt/month)
Routes	Australia–Japan	Brazil–Japan	Suriname–USA	Guinea–USA	1972 ship
Nautical miles	3600	11 500	4500		
Vessel size					
15 000 dwt			4.00	6.30	6.60
25 000 dwt	2.60	6.80	3.30	4.80	5.23
35 000 dwt			3.10	4.30	3.72
60 000 dwt	2.10	5.40	3.00	3.90	—
100–110 000 dwt	1.80	4.00	—	—	—
120–130 000 dwt	—	3.50	—	—	—
140–150 000 dwt	—	3.40	—	—	—

Sources:
[a] United Nations, Maritime Transportation of Iron Ore, TD/B/C.4/105 Rev. 1, UNCTAD, p. 92.
[b] Brandon, J., Haclin, G. and Rowbotham, P., Transport of Bauxite and Alumina Volume, Costs, Technical Background, Future Trends, ID/W.G. 273/10, UNIDO, 1978.
[c] Figures are capital and operating costs from H. P. Drewry, Shipping Consultants, The Operation of Dry Bulk Shipping Present and Prospective Trading Costs in the Context of Current and Future Market Trends, London, 1979, p. 33.

that they will employ their own vessels.

Transport costs are a function of distance, efficiency of transport systems (port capacity, cargo handling, vessel size and type) and organization (types of contracts, ownership). The economics of shipping can be simply stated: freight costs increase with the length of haul and decrease with increased vessel size. It is important to note that freight or transport *costs* are not the same as freight *rates*. Published freight rates are usually for spot market transactions or single voyages and so apply to only a small fraction of tonnage shipped. Such freight rates are volatile and reflect demand and supply in shipping. As such, they are poor indicators of the underlying costs of transport. Consequently, the best method to examine transport costs is to look at the spread between c.i.f. and f.o.b. prices for commodities, as for example the c.i.f. and f.o.b. iron ore from Brazil to the United States, bauxite and alumina from Suriname and Guinea to the United States, or copper from Peru and the Philippines to the USA. Comparison of the spread between the, c.i.f. and f.o.b. prices indicates the share of transport costs in c.i.f. values. This share is highest for iron ore and bauxite (20–29%) and lowest for alumina and copper ores and concentrates (4–9%).

Studies of transport costs often include estimates of capital charges, operating costs (administration, crewing, maintenance and repair) and voyage costs (fuel costs, port charges) as well as a calculation for backhauling (which can reduce voyage costs by 50%). Once the magnitude of transport costs are estimated, their incidence — that is, who bears the cost, producer or consumer — can be established. Since ore minerals are relatively price-inelastic in terms of supply, because of the long development period required to open new mines or expand old ones, and the tendency to accumulate stocks in the short-term, the producers of iron ore, bauxite and alumina will bear any increase in transport costs. As a general rule, c.i.f. prices tend to be equal in an international market, and normal freight differentials will have to be absorbed by the distant sellers in the form of a lower f.o.b. return than that received by nearby sellers.

For example, if Japanese steel makers have decided on a c.i.f. price of $25–30 per ton for iron ore delivered from Australia, they will require that Brazilian iron ore be delivered at the same price even though there is an 8000 mile difference in transport. Consequently, Brazilian iron ore producers will be forced to lower their f.o.b. price relative to the Australian.

One important factor which influences transport costs is the price of bunker fuel. The following table illustrates the trend in bunker prices.

TABLE 55
Bunker prices 1972–78 (US $ per ton)

	Fuel oil	Marine diesel
1972	11	15
1973	26	36
1974	85	110
1975	72	98
1976	78	106
1977	85	120
1978	80	125
1979[a]	129	247
1980[a] (first 6 months)	160	319
1980[a] (October)	215	

Sources: 1972–1978 from H. P. Drewry, Shipping Consultants, *Operation of Dry Bulk Shipping: Present and Prospective Trading Costs in Context of Current and Future Marketing Trends*, 1979, p. 19.
[a] 1979–80 figures are simple averages of spot bunkers at Rotterdam, Ras Tanura, US Gulf and Singapore. It should be noted that US Gulf prices were substantially below the other three and so pulled down the average – taken from *Lloyds Shipping Economist*, June 1980.

Of interest is the period 1978–80 when fuel oil increased 2.5 times. Freight rates did not increase to a comparable extent because of excess capacity in shipping. What increase did occur was partly absorbed by the producing countries.

If the f.o.b. price of raw materials remains constant (in real terms), there is an incentive for developing countries to process minerals in order to lessen the depressing effect of rising transport costs on the net prices of the minerals they produce. However, where transport costs are a small share of the c.i.f. value of the minerals produced, such an incentive is weak.

5.6 BY-PRODUCTS

Among the other factors which influence the feasibility of further processing is the production of by-products which can be marketed. For

TABLE 56
Relative importance of transport costs

Smelter inputs	Base operating costs	250% increase[a] in transport costs (due to 250% increase in price of fuel oil)	250% increase in energy costs in smelting
1.93 tons alumina $240 + $10 transport charge (250% = $25)	483	511	
Electricity 14 000 kWh × 2.4 cents per kWh (250% = 6¢ per kWh)	336		840
0.1 fuel oil, $150 per ton	15	37	
Electrolyte (0.06 tons at $450) Electrodes (0.6 tons at $150) Labour and supervision 15 man hours (at $6 per man hour) Miscellaneous	232		
	1066	1116 (5% increase)	1570 (47% increase)

[a] The example makes the extreme assumption that there is no excess capacity in shipping and that fuel costs are 100% of transport costs so that any energy cost increase is fully reflected in the rise in transport costs. In reality there is now excess capacity and bunker fuel costs are only 50% of total transport costs.

example a by-product of copper smelting is sulphuric acid, which is costly to transport. If local markets are insufficient to absorb it, the profitability of the smelting operation can be jeopardized. A market for sulphuric acid will allow the Philippines to process their concentrates whereas the lack of such a market remains an obstacle to processing in British Columbia, as has been noted earlier.

5.7 ENERGY REQUIREMENT AND AVAILABILITY

In 1979, developing countries produced almost 31% of world primary commercial energy, but there was a disparity between production in member countries of the Organization of Petroleum Exporting Countries (OPEC) and non-OPEC countries, the latter accounting for only 6.54% of total primary commercial production. Table 57 illustrates these differences.

A recent UNIDO report estimates that developing countries currently consume only 13%

of the world's commercial energy.[97] World Bank studies[98] project that energy demand in developing countries may increase at a faster rate than that of developed countries for the following reasons:

(a) developing countries may maintain higher growth rates;

(b) continued industrialization and urbanization may make these economies more energy-intensive;

(c) rising income levels may increase the demand for energy-consuming amenities;

(d) commercial energy may be substituted for non-commercial energy.

The primary metal sector, in particular iron and steel, aluminium and copper, is a substantial energy consumer within industry. Countries should avoid establishing energy-intensive industries if they would have to import both the raw materials and the energy. Energy surplus countries, on the other hand, should be able to establish energy-intensive industries provided the processed metals can be exported.

TABLE 57
Production of primary commercial energy in 1979 (thousands of tons oil equivalent)

	Total primary energy	Solids (coal, lignite)	Liquids	Natural gas	Electricity
World	6 575 153	1 882 707	3 227 318	1 267 657	197 476
Developing countries	2 041 625	84 017	1 809 735	118 362	29 516
	(31.1%)	(4.5%)	(56.1%)	(9.3%)	(14.9%)
OPEC	1 613 498	1014	1 550 778	59 640	2066
	(24.6%)	(0.1%)	(48.1%)	(4.7%)	(1.0%)
Non-OPEC	428 127	83 003	258 957	58 722	27 450
	(6.5%)	(4.4%)	(8.0%)	(4.6%)	(13.9%)

Source: United Nations, Yearbook of World Energy Statistics, 1979, New York, 1981.

NOTES

[1] Wall, D. 1980. Industrial processing of natural resources. *World Develop.* **7**, 303–316.

[2] United Nations General Assembly, Resolution 3202 (S-VI), Sec. 1 (1), para. (3); UNCTAD, Trade and Development Board, Resolution 124; UNIDO (ID/CONF. 3/31), *Report of the Second General Conference*, Chapter 4, Lima Declaration and Plan of Action, para. 59 (D).

[3] United Nations Industrial Development Organization, *Industrialization in Latin America*, Declaration of the Latin American Conference on Industrialization, Mexico City, December 1974.

[4] For a discussion of the problem of defining processing activities, *see* Wall, D. 1980. Industrial processing of natural resources. *World Develop.* **7**, 303–316.

[5] The particular case of copper rod is discussed by Mezger, D. 1975. The European copper industry and its implications for the copper-exporting developing countries. *In* A. Seidman (ed.) *Natural Resources and National Welfare: The Case of Copper.* Praeger, New York.

[6] United Nations Industrial Development Organization 1980. *Mineral Processing in Developing Countries* p. 84; World Bank 1975. *Export-Oriented Processing of Primary Commodities in Developing Countries.* p. 84.

[7] Brown, M. 1979. *The Location of Copper Processing: Some Preliminary Notes*, Organization for Economic Co-operation and Development, Paris; Commodities Research Unit, 1975. *Study of the Degree and Scope for Increased Processing of Primary Commodities in Developing Countries*, prepared for the United Nations Conference on Trade and Development, Geneva. Hashimoto, H. 1979 *Bauxite Processing in Developing Countries: A Preliminary Review of Issues and Prospects.* World Bank, Washington; le Moal, Y. 1979. *Report on Economics of Processing Iron Ore to Steel*, Organization for Economic Co-operation and Development, Paris; Roemer, M. 1979. Resource-based industrialization in developing countries. *J. Develop. Econ.*, **6**, 163–202; UNCTAD V, 1979. *The Processing Before Export of Primary Commodities: Areas for Further International Co-operation* (TD/229/Supp. 2), Manila 1979; UNCTAD *Measures to Expand Processing of Primary Commodities in Developing Countries* (TD/B/C.1/197), Geneva; UNCTAD, *Processing of Primary Products in Developing Countries: Problems and Prospects* (MD/79), Geneva; UNIDO, 1971. Mineral processing in developing countries. *In: Industry 2000: New Perspectives.* Collected Background Papers, Vienna, Vol. 5; World Bank, 1979. *Export Oriented Processing of Primary Commodities in Developing Countries.* Commodity Note No. 14, World Bank, Washington.

[8] Economic Commission for Europe, 1976. *Increased Energy Economy and Efficiency in the ECE Region.* (E/ECE/883, Rev. 1), Geneva, p. 14.

[9] United Nations Industrial Development Organization, 1980. *Mineral Processing in Developing Countries.* United Nations, New York.

[10] United Nations General Assembly resolution 3362 (S-VII), *Development and International Economic Co-operation*, 16 September 1975; UNCTAD, 1976. *New Directions and New Structures for Trade and Development* (TD/183), pp. 19–23.

[11] *See* Myint, H. 1971. *Southeast Asia's Economy in the 1970s.* Praeger, New York.

[12] Rweyemamu, J. 1973. *Underdevelopment and Industrialization in Tanzania: a Study of Perverse Capitalist Industrial Development.* Oxford University Press, Thomas, C. *Dependence and Transformation: The Politics of the Transition to Socialism.* Monthly Review Press, New York.

[13] Mikesell, R. (ed.) 1971. *Foreign Investment in the Petroleum and Mineral Industries.* Johns Hopkins University Press, Baltimore; Mikesell, 1970. United States private investment in the extractive industries of the developing countries: changing attitudes and patterns. *In:* I. A. Litvak and C. J. Maule (eds): *Foreign Investment: the Experience of Host Countries.* Praeger, New York, pp. 355–381.

[14] Hassan, M. F. 1975. *Economic Growth and Employment Problems in Venezuela: an Analysis of an Oil-Based Economy.* Praeger, New York; Knauerhase, R. 1975. *The Saudi Arabian Economy.* Praeger, New York; F. Fesharaki and D. T. Isaak, 1981. *The Emerging Petroleum Product Market: Implications for World Refining and Transport Industries.* Resource Systems Institute, East–West Center, Honolulu, June 1981.

[15] *See* Organization for Economic Co-operation and Development, 1979. *Information Note: Research on Processing of Natural Resources*. DC/F(79)1503, OECD, Paris, p. 16.

[16] Amin, S. 1974. *Accumulation on a World Scale*. Monthly Review Press, New York; F. Bonilla and R. Girling (eds), 1973. *Structures of Dependence*, Stanford University Press, Stanford; Frank, A. G. 1969. *Capitalism and Underdevelopment in Latin America*. Monthly Review Press, New York; Furtado, C. 1964. *Development and Underdevelopment*. University of California Press, Berkeley; Rweyemamu, J. 1973. *Underdevelopment and Industrialization in Tanzania: a Study of Perverse Capitalist Industrial Development*. Oxford University Press, Nairobi; Sunkel, O. 1968. National development policy and external dependence in Latin America. *J. Develop. Stud.*, **6** (October).

[17] Roemer, M. 1977. *Resource-Based Industrialization in the Developing Countries*. Develop. Discussion Pap. No. 21, Harvard Institute for International Development, Cambridge MA, pp. 57–58.

[18] Mezger, D. 1975. The European copper industry, *op. cit.*

[19] Della Valle, P. 1975. Productivity and employment in the copper and aluminium industries. *In:* A. S. Bhalla (ed.) *Technology and Employment in Industry*. International Labour Office, Geneva, pp. 272–308.

[20] For documentation of this pattern in the case of Jamaica, *see* Tidrick, G. M. 1975. Wage spillover and unemployment in a wage-gap economy: the Jamaican case. *Econ. Develop. Cultural Change*, **23** (January) 306–324.

[21] United Nations Centre on Transnational Corporations 1981. *Transnational Corporations in the Bauxite/ Aluminium Industry*. United Nations, New York, p. 1.

[22] Stigler, G. 1968, *The Organization of Industry*. Homewood, Illinois, Irvin, ch. 13; Radetzki, M. 1977. Where should developing countries' minerals be processed? *World Develop.*, **4**, 331.

[23] Radetzki, 1977, *ibid.*, p. 332.

[24] Mikdashi, Z. 1976. *The International Politics of Natural Resources*. Cornell University Press, Ithaca, N.Y., pp. 27–31.

[25] For a more detailed explanation of this point, *see* Zorn, S. 1980. Recent trends in LDC mining agreements. *In:* S. Sideri and S. Johns (eds) *Mining for Development in the Third World*. Pergamon, New York, pp. 210–228.

[26] *See* Wall, 1980. Industrial processing of natural resources, pp. 305–306.

[27] *Ibid.*

[28] G. Helleiner and D. Welwood, 1978. *Raw Material Processing in Developing Countries and Reductions in the Canadian Tariff*. Discussion Pap. No. 111, Economics Council of Canada, April 1978, p. 34.

[29] Walter, I. 1971. Non-tariff barriers and the export performance of developing countries. *Amer. Econ. Rev.* (May).

[30] UNIDO, 1971. *Mineral Processing in Developing Countries*, pp. 10, 24, 51.

[31] Such practices are described in detail in *Report of the Intergovernmental Group of Experts on a Code of Conduct on Transfer of Technology*, UNCTAD, document TD/B/C.6/1, Annex 3.

[32] Mikesell, R. 1979, *The World Copper Industry*. Johns Hopkins University Press, Baltimore, pp. 103–110; Herfindahl, O. C. 1959. *Copper Costs and Prices 1870–1957*. Johns Hopkins University Press, Baltimore.

[33] UNCTAD, *op. cit.*, note 27, p. 10.

[34] *Ibid.*, p. 9.

[35] *See* UNIDO, 1971. *Mineral Processing in Developing Countries*, p. 113.

[36] *See* Teece, B. J. 1977. Technology transfer by multinational firms: the resource costs of transferring technical know-how. *Econ. J.*, **87** (346).

[37] UNIDO, 1971. *Mineral Processing in Developing Countries*, pp. 131–133.

[38] Mezger, D. 1975. The European copper industry and its implications for the copper-exporting underdeveloped countries, pp. 71-74.

[39] Takeuchi, K. 1977. *The Potential for Increased Processing of Nonfuel Minerals in Developing Countries*. World Bank, Washington, p. 8 (mimeo).

[40] OECD, *Information Note: Research on Processing of Natural Resources*, pp. 12–13.

[41] UNIDO, 1980. *Industrial Processing of Natural Resources*. UNIDO, Vienna (mimeo) p. 15; Roemer, *Resource-Based Industrialization in the Developing Countries: a Survey of the Literature*, pp. 27–28.

[42] Roemer, 1977. *Resource-Based Industrialization in the Developing Countries*, p. 28; Brubaker, S. 1967. *Trends in the World Aluminum Industry*. Johns Hopkins University Press, Baltimore, p. 232; Sartorius, P. and Henle, H. 1968, *Forestry and Economic Development*. Praeger, New York, p. 114.

[43] UNIDO, 1980. *Industrial Processing of Natural Resources*, p. 41.

[44] Recent freight rates quoted for bauxite and alumina from Australia to the West Coast of the United States are in the range of $13–14 per ton (*Industrial Minerals*, July 1981, p. 53).

[45] Roemer, 1977 *Resource-Based Industrialization in the Developing Countries*, p. 53.

[46] Ohlin, B., Hesselborn, P. O. and Wijkman, P. M. (eds) 1977. *The International Allocation of Economic Activity*. Holmes and Meier, New York, p. 276.

[47] *Copper Studies*, 1977. (May 23) p. 1.

[48] *International Legal Materials*, 1980. **19** 524–525.

[49] Radetzki, Where should developing countries' minerals be processed?, *op. cit.*, p. 328.

[50] Mikesell, R. 1975. *Foreign Investment in Copper Mining*. Johns Hopkins University Press, Baltimore, p. 121.

[51] On this topic, *see* Knop, H. (ed.) 1977. *The Bratsk–Ilimsk Territorial Production Complex*. International Institute for Applied Systems Analysis, Laxenburg.

[52] Totopoulos, P. and Nugent, J. 1973. A balanced-growth version of the linkage hypothesis: a test. *Q. J. Econ.*, **87**, (2) 157–71.

[53] United Nations Centre on Transnational Corporations, 1981, *Transnational Corporations in the Bauxite-Aluminium Industry*. E.81.II.A.5, UN, New York, p. 1.

[54] *Metal Bulletin*, 1977. World aluminium survey. Metal Bulletin Ltd., Surrey, U.K.

[55] World Bank, 1982. *Price Prospects for Major Primary Commodities*. (August).

[56] *Financial Times*, 1980. Vying for Growth vs. Steel and Plastics. *Financial Times*, London (October 8).

[57] Hashimoto, H. 1979. *Bauxite Processing in Developing Countries: A Preliminary Review of Issues and Prospects*. World Bank Discussion Pap., November.

[58] United Nations Industrial Development Organization, *Final Report Workshop on Case Studies of Aluminium Smelters*, p. 47.

[59] UNIDO, *ibid*, p. 21.

[60] Because demand was at a very low level at the end of 1981, actual transaction prices were much lower (probably around 60 US cents/lb) since all major producers granted substantial discounts. The LME price was lower again, at little more than 50 US cents/lb, though only a very limited amount of aluminium is sold on the basis of the LME price.

[61] Battelle Columbus Laboratories, *Energy Use Patterns in Metallurgical and Non-Metallic Mineral Processing*, June 1975, pp. 6–7.

[62] *Engineering and Mining Journal*, New high-intensity aluminium pots cut costs and spark wide interest, July 1980, p. 35.

[63] Charpie, R. A. and MacAvoy, P. W. 1978. Conserving energy in the production of aluminium. *Resour. Energy*, 34.

[64] Elliott-Jones, M. E. 1974. Aluminium. *In: Energy Consumption in Manufacturing*, National Conference Board, p. 533.

[65] UNIDO, *Mineral Processing in Developing Countries*, *op. cit.*, p. 77.

[66] Perlman, R. 1980. Continuous casting of copper wirerod-prospects for the world copper industry. *CIPEC Q. Rev.* (January-March).

[67] Davies, M. H. 1979. Future strategy: the case of copper. *CIPEC Q. Rev.* (July–September) p. 27.

[68] World Bank 1980. *Price Prospects for Major Primary Commodities*. Rep. No. 814/80, World Bank, p. 291.

[69] Etheridge, W. 1980. *Demand for metals*. Pap. Joint Meet. Inst. Min. Metall. and Soc. Min. Eng. (AIME) London, May 1980.

[70] Davies, *op. cit.*, p. 28.

[71] Hydrometallurgical processes have been used for treatment of oxide ores for many years and are not considered here. The techniques mentioned here have been developed for sulphide concentrates.

[72] Gelb, B. and Pliskin, J. 1979. *Energy Use in Mining: Patterns and Prospects*. Ballinger, Cambridge, p. 61.

[73] Battelle Columbus Laboratories 1975. *Energy Use Patterns in Metallurgical and Non-Metallic Mineral Processing*. p. 46.

[74] Battelle Columbus Laboratories, 1975. *Energy Use Patterns in Metallurgical and Non-Metallic Mineral Processing*.

[75] King, D. E. C. 1980. Current and developing treatment technology and cost comparison with regard to capital, labour and energy costs. *In: Copper in British Columbia*. B.C. Department of Mines.

[76] Drewry, H. P. (Shipping Consultants) 1979. *The Prospects for Seaborne Iron Ore Trade and Transportation*. London, p. 25.

[77] Gamboa, A. 1979. *The Iron and Steel Industry Development in the Arab Countries*. Iron Steel Mag. (July) 8–10.

[78] Economic Commission for Europe, 1980 Data Bank.

[79] World Bank, 1982. *Price Prospects for Major Primary Commodities*.

[80] UNCTAD, 1979. *The World Commodity Situation and Outlook*. (TD/B/C.1/207).

[81] World Bank, 1982. *Price Prospects for Major Primary Commodities*.

[82] Field, L. 1979. The impact of energy conservation. *Iron Steel Mag.* (January) p. 10.

[83] For a discussion of the various measures of capital intensity and their deficiencies *see* Bhalla, A. S. 1975. The concept and measurement of labour intensity. *In:* A. S. Bhalla (ed.) *Technology and Employment in Industry*. International Labour Organization, Geneva.

[84] This method was followed in Economic Commission for Europe, UN, 1977. *Structure and Change in European Industry*. United Nations, New York.

[85] McKern, R. B. 1977. A survey of opportunities for low scale manufacture in developing countries. Work. Pap. for OECD Expert Meet. on *Downscaling and Adaptation of Industrial Technology*, Paris, June 1977. One exception to this has occurred in the iron and steel industry where Mexico perfected the direct reduction process which is less capital-intensive than the conventional blast furnace process route.

[86] Sir Mark Turner, 1980. *The Investment Problems of Producer Companies and Countries*. Joint IMM–AIME Meet., London, May 29, 1980. The $8000 estimate for copper may be too low. The capital cost of mine capacity at Andacollo (Chile) has been estimated at $6000 per year (*American Metal Market*, 18 September 1980) and this study estimated that new smelting/refining capacity would cost at least $2500 per ton. Charles Barber of Asarco has estimated that the Cuajone mine complex would have cost $10 800 per ton of capacity if it had been brought on stream in 1980.

[87] Phiri, D. A. R. 1980. *Investment in Developing Countries: A Matter of Give and Take*. Joint IMM–AIME Meeting, London, May 1980.

[88] The term 'project financing' is not used in the usual sense, that is a combination of non-recourse loans supported *only* by the cash flow of the project. In any case that type of financing is becoming exceedingly rare as commercial lenders are asking for loan guarantees from the parent company of the TNC involved or from the government of the developing country.

[89] Tugendhat, C. *Raw Materials, The Third World and the European Community*.

[90] Barber, C. 1980. *Mineral Investment in an Anxious World*. Joint IMM–AIME Meeting, London, May.

[91] Radetzki, M. and Zorn, S. 1979. *Financing Mining Projects*. Mining Journal Books, London, pp. 84–103.

[92] An exception to this general statement is the fact that some oil-exporting countries, Iran, Bahrain and Dubai, have financed aluminium smelters in their own countries.

[93] *See* the correlations between skill ratios and capital per man computed by Hufbauer, G. C. 1970. The impact of national characteristics and technology on the commodity composition of trade in manufacture goods. *In:* R. Vernon (ed.) *The Technology Factor in International Trade*. Columbia University Press, New York. Balassa measured skill intensity using industry wage rates and found it is closely correlated with physical capital intensity for resource-based industries reinforcing the conclusion that less developed countries would not have a comparative advantage in these products (*see* Balassa, B. 1977. A *'Stages'* Approach to Comparative Advantage. World Bank Staff Working Pap. No. 256, 1977).

[94] *Engineering and Mining Journal*, 1980 (September), p. 65.

[95] UNIDO, 1977. *Final Report of the Workshop on Case Studies of Aluminium Smelter Construction in Developing Countries*. ID/WG 250/18, Vienna.

[96] It is estimated that 95% of iron ore is shipped under long-term arrangements. *See* Drewry, H. P. (Shipping Consultants) 1977. *The Nature of Dry Bulk Shipping*. Seminar on Shipping Costs and Revenues, London.

[97] UNIDO, 1980. *Energy Intensity and Industrial Development Strategy*. Paper prepared by UNIDO for the ACC Task Force on Long-Term Development Objectives, New York, May 22–28, 1980.

[98] World Bank, 1979. *Energy Options and Policy Issues in Developing Countries*. Staff Work. Pap. No. 350.